Introduction to OBD II

Roy S. Cox
Student Edition

DELMAR
CENGAGE Learning

Australia • Canada • Mexico • Singapore • Spain • United Kingdom • United States

Introduction to OBD II

Roy Cox

Vice President, Technology
and Trades SBU:

Alar Elken

Editorial Director:

Sandy Clark

Product Development Manager:

Kristen Davis

Marketing Director:

Beth A. Lutz

Marketing Specialist:

Brian McGrath

Marketing Coordinator:

Marissa Maiella

Production Director:

Mary Ellen Black

Development Editor:

Kim Blakey

Senior Production Manager:

Larry Main

Production Editor:

Benj Gleeksman

Art/Design Coordinator:

Benj Gleeksman

For product information and technology assistance, contact us at
Cengage Learning Customer & Sales Support, 1-800-354-9706

For permission to use material from this text or product,
submit all requests online at **cengage.com/permissions**
Further permissions questions can be emailed to
permissionrequest@cengage.com

ISBN-13: 978-1-4180-1220-5

ISBN-10: 1-4180-1220-3

Delmar Cengage Learning
5 Maxwell Drive
Clifton Park, NY 12065-2919
USA

Cengage Learning products are represented in Canada by Nelson Education, Ltd.

For your lifelong learning solutions, visit **delmar.cengage.com**

Visit our corporate website at **www.cengage.com**

Notice to the Reader

Publisher does not warrant or guarantee any of the products described herein or perform any independent analysis in connection with any of the product information contained herein. Publisher does not assume, and expressly disclaims, any obligation to obtain and include information other than that provided to it by the manufacturer. The reader is expressly warned to consider and adopt all safety precautions that might be indicated by the activities described herein and to avoid all potential hazards. By following the instructions contained herein, the reader willingly assumes all risks in connection with such instructions. The publisher makes no representations or warranties of any kind, including but not limited to, the warranties of fitness for particular purpose or merchantability, nor are any such representations implied with respect to the material set forth herein, and the publisher takes no responsibility with respect to such material. The publisher shall not be liable for any special, consequential, or exemplary damages resulting, in whole or part, from the readers' use of, or reliance upon, this material.

Printed in the United States of America
2 3 4 5 6 7 11 10 09 08

Contents

Preface

Introduction

This book is about OBD II, the standard diagnostic system found on all vehicles sold in the United States since the 1996 model year. Both entry-level and experienced technicians will find useful information about how this sophisticated system works and learn effective diagnostic and troubleshooting tips, tricks, and techniques.

Huge amounts of information about OBD II and engine management systems are readily available. As a working automobile service technician, it is likely that you have already been exposed to training programs and read lots of diagnostic routines. However, much of the information out there involves specific diagnostic steps or electronic theory that really doesn't help the technician understand what the tests mean or how OBD II interrelates to the engine control functions and emission control systems. This book was written to unravel the tangle of details and provide a commonsense approach to explain the system relationships and provide the technician with easy-to-understand information that can streamline day-to-day driveability service.

The OBD II system is exactly what its name implies, an on-board diagnostic system. It is an accessory system, which is designed to monitor the efficiency of the engine and its subsystems. Its purpose is to identify and report malfunctions that increase exhaust and/or evaporative emissions. In effect, OBD II is like having a portable emissions tester on-board, which is turned on whenever the vehicle is being driven. To be sure, it is related to the engine and powertrain controls, but it is *not* an engine management system. OBD II interfaces with the engine management computer as it "watches" the operation of the fuel system, ignition system, emission controls, and the engine itself. The OBD II software is even located inside the computer. However, it is most important to remember that OBD II does not replace the computer, or any other controls. While problems identified by OBD II can often cause driveability symptoms, it is specifically looking for problems that will increase exhaust emissions.

Through the course of this book, we will explore the history of this sophisticated and fascinating diagnostic system, learn how it functions, and look into the inter-relationships that exist between OBD II and engine management. Best of all, we will learn to interact comfortably with OBD II. We will learn its language and the huge amount of data and information it provides to aid the auto repair technician in troubleshooting and repairs. In addition, we will learn how to run our own tests to troubleshoot driveability symptoms quickly and easily. We will discuss troubleshooting techniques, both with and without the use of expensive test equipment or exhaust analyzers.

As the author of this course, I know that you will enjoy learning the finer points of OBD II. More importantly, I know that you will learn lots of tips and techniques that you can use everyday when troubleshooting and repairing problems in today's complex automotive systems. My promise to you is to make your learning experience as beneficial as possible by making the contents of this course easy to understand and fun to learn.

Sincerely,
Roy S. Cox

How to Use This Book

The best way to approach this or any other learning project is to enter into it with an open mind. As you read the material in each chapter, look over the information that is being discussed and make sure that you are comfortable with the terms, concepts, and tests. Each chapter is a building block that you need to feel confident in using as you move into the next chapter and add more knowledge. By progressing in this manner, you will build a thorough understanding of the OBD II system and the engine management systems it is related to. In addition, you will accumulate a series of diagnostic and service aids to make your drive-ability jobs easier and your test routines quicker and smoother.

This knowledge base is what you are intended to take away from the effort you put into learning, and it is the single most important tool you can obtain. If you know what is supposed to happen, when it should happen and why, you have a great advantage in diagnosing and repairing virtually any part of a motor vehicle.

The book is organized in a logical order of progress. At the end of each chapter, you will find a mini-quiz of review and reinforcement questions. Some of them may require you to use your knowledge of how the systems and components function rather than simply going back through the chapter and looking up the answer. As you work with the review and reinforcement questions, do not try to simply memorize the correct answers. Instead, try to expand your horizons of knowledge. If you do not know why a particular answer choice is right or wrong, find out! In this way, you will be able to transfer this knowledge to any other questions you may encounter on the same subject area.

As a further means of reinforcement, practice the tests and troubleshooting procedures. Familiarize yourself with how the tests are performed, what the results mean, and why the routines are laid out in a particular order. Most importantly, become comfortable with the use of a variety of diagnostic equipment and learn how to get the most benefit from it. Your instructor has been supplied with suggested shop exercises to help you learn. In addition, you will have the opportunity to perform practice sessions on the Interactive Troubleshooting CD-ROM Tool, which is specially designed to accompany this course.

Now, let's proceed into the exciting system known as OBD II.

1

The Evolution of On-Board Diagnostics

The sophisticated self-diagnostic capabilities of today's modern automobiles are a far cry from those of systems used in the early years of emission controls. Even the last 10 years have seen sweeping changes in the operation and problem diagnosis of ignition, fuel delivery, and emission control systems, and the evolution continues. Each year, more detailed information is automatically monitored and controlled by on-board processors, including everything from the engine's speed and load to the audio system volume. The story of these complex systems and how they operate is an interesting one.

Milestones in the History of On-Board Diagnostics

Although **on-board diagnostics (OBD)** first made its appearance in the late 1970s, its story began in 1960, in the state of California. The California Bureau of Air Sanitation first began to address air pollution in metropolitan areas in the early 1950s, but nothing was done to address or improve pollutants generated by

Figure 1–1   The PCV valve, still present on many new models, was the first emission control device.

the automobile. In 1960, the Motor Vehicle Pollution Control Board (MVPCB) was formed to regulate automotive pollutant emissions. Four years later, the first emission control device was mandated for 1966 model year vehicles to be sold in the state. That device, still present on most engines today, is the **positive crankcase ventilation (PCV) valve** (Figure 1–1). The PCV system captures unburned air/fuel mixture that escapes past the piston rings and returns it to the intake system where it is burned, rather than being allowed to escape into the atmosphere.

California has been, and continues to be, at the forefront of air pollution awareness and emission regulations, both automotive and industrial.

In 1968, the United States Congress enacted the first federal automotive emissions legislation requiring PCV systems on all 1968 and later model year vehicles sold in the United States. So began the evolution of automotive emission regulations, which were the original driving force behind the many electronic controls and monitors we have today.

Also in 1968, the California Air Resources Board (CARB), which encompassed the duties of both the MVPCB and the Bureau of Air Sanitation, was established to oversee and regulate all air pollution in the state. CARB established test procedures and regulations to ensure cleaner running vehicles that were soon to be emulated by Congress in the form of federal emission legislation.

In 1970, Congress passed the first national Clean Air Act, which called for a 90 percent reduction in motor vehicle exhaust emissions. The Clean Air Act began by targeting **carbon monoxide (CO)** and **hydrocarbon (HC)** emissions and set standards that were to be applied to 1975 model year vehicles sold in the United States. **Oxides of nitrogen (NO$_x$)** emissions were targeted for reduction beginning with 1976 model year vehicles. The U.S. Environmental Protection Agency (EPA) was established by Congress and was charged at the federal level with the same duties and responsibilities as CARB had assumed in California.

The Clean Air Act of 1970 was the beginning of a series of regulations that called for ever-decreasing limits for automotive exhaust emissions. As auto manufacturers struggled to meet the mandated limits, a series of new emission controls and redesigned components was introduced, including the **exhaust gas recirculation (EGR)** system, the charcoal canister vapor recovery system, and the *catalytic converter*. Even with these new additions, it was very difficult to meet the standards with carburetors and mechanically controlled ignition timing. In time, this series of events led the automotive manufacturers to the need for precise monitoring and control of fuel delivery, ignition, and emission control systems.

By 1981, new vehicles sold in the United States were equipped with *three-way catalytic converters*, *oxygen sensors*, and *closed-loop engine management systems* that required the use of on-board computers. In order to keep these systems operating at peak efficiency, some manufacturers began building their on-board computers with the capability to test, or monitor, the operation of the various components that managed the engine's fuel delivery, ignition and emission controls. OBD was born!

History of OBD I and II

From 1981 through 1987, the Society of Automotive Engineers (SAE), the EPA, and CARB were working toward standardizing and regulating automotive emission controls. In 1988, OBD I compliance for new vehicles sold in California was enacted by CARB. The OBD I standards required a certain universal level of self-diagnostics and monitoring built into all engine control computers, as well as a system of trouble codes to guide a repair technician in finding and fixing a given problem. However, manufacturers were free to design their own means of accessing the diagnostic information and to provide their own individual trouble code definitions and troubleshooting procedures.

As might be imagined, a wide variety of methods were implemented and little standardization was accomplished. Some manufacturers provided substantially more diagnostic information to the technician than others. Some models could communicate stored trouble codes and a live data stream for numerous sensors and components to a *scan tool*, so the technician could get an accurate picture of what the on-board computer was "seeing." Other models provided only "flash codes," trouble code numbers that were read by counting the number of times the **malfunction indicator lamp (MIL)** flashed after entering the system's diagnostic mode. This was usually done by jumping two wires together on the vehicle's

assembly line diagnostic link (ALDL), now called the **diagnostic link connector (DLC)**. DLCs were of many sizes and shapes, which required a wide variety of connector plugs to be supplied with scan tools.

Even the method of displaying flash codes was not standardized. For example, most General Motors vehicles flashed each code three times and always displayed code "12," no RPM signal, to verify that the self-diagnostic mode was working. Most Chrysler products flashed each code once, then flashed "55" to indicate the end of the diagnostic message. Ford Motor Company vehicles usually flashed two series of codes, separated by a single flash of the MIL, indicating the number "10." The first series were current problems, present as the codes were being read. The second series were codes stored in memory, but not currently present. Some Asian and European imports contained **light emitting diodes (LEDs)** on the side of their engine computers, where the trouble codes were displayed.

Needless to say, a service manual showing how to retrieve the trouble codes and data and what the trouble codes meant for each year, make, and model was absolutely essential before any accurate troubleshooting and repair could take place.

Figures 1–2 and 1–3 show trouble code definitions for two specific vehicle models equipped with OBD I. As you compare the two charts, you will note that like trouble code numbers almost never mean the same thing and that one model has significantly more trouble codes than the other. Variance in the number of codes used is not unusual, as manufacturers tend to enhance their engine management systems and diagnostic capabilities when new or redesigned models are introduced. In Figures 1–2 and 1–3, the Dodge Dakota was not significantly redesigned from 1991 until 1996 while the 1993 Chevrolet Camaro was a new design for that model year.

A new set of regulations, which were later called OBD II, were proposed by CARB in 1988 and adopted in 1989. The OBD II standards called for a single universal method of retrieving trouble codes and specified which data and self-diagnostic functions had to be readable on all models. Universal trouble code definitions and acronyms also were mandated. So, the same component always had the same name and the same trouble code number always meant the same thing! The manufacturers were still free to add more functions and use special, proprietary diagnostic tools and methods, but the minimum OBD II requirements had to be accessible to all repair technicians equipped with a universal OBD II–compliant scan tool. OBD II legislation set target dates beginning with the 1990 model year and called for 100 percent compliance by the 1996 model year.

The OBD II regulations, complete with required 1996 model year compliance, were enacted as federal law by Congress as part of the sweeping 1990 amendments to the Clean Air Act. These changes resulted in tightened standards for automotive exhaust emissions and addressed numerous emission concerns in greater detail than ever before.

Code Number	Definition	MIL On?
11	No crank position signal	Yes
12	Battery disconnect/power loss	No
13	No change in manifold absolute pressure from start to run	Yes
14	Manifold absolute pressure sensor voltage too low	Yes
15	No vehicle speed sensor signal	Yes
17	Engine is cold too long	No
21	Oxygen sensor stays at center or shorted to voltage	Yes
22	Coolant temp. sensor voltage too high or too low	Yes
23	Intake air temp. sensor voltage too high or too low	Yes
24	Throttle position sensor voltage too high or too low	Yes
25	Idle air control motor circuits	Yes
27	Fuel injector control circuit (any cylinder)	No
31	Evaporative emission canister solenoid circuit	Yes
32	EGR solenoid circuit or EGR system failure	Yes
33	A/C clutch relay circuit	No
34	Cruise control solenoid circuit or switch	No
35	Radiator fan relay control circuits	No
41	Alternator field not switching properly	Yes
42	Auto shut down relay control circuit	Yes
44	Battery temp. sensor out of limits	No
46	Charging system voltage too high	Yes
47	Charging system voltage too low	Yes
51	Oxygen sensor signal stays lean	Yes
52	Oxygen sensor signal stays rich	Yes
53	Internal PCM failure	No
55	Completion of diagnostic codes display	No
62	Service reminder indicator miles not stored	No
63	PCM EEPROM writing (programming) failure	No

On this vehicle model, the Malfunction Indicator Lamp displays "Check Engine."

Figure 1–2   Trouble Code Definitions: 1995 Dodge Dakota With 2.5L 4 Cylinder Engine.

Code Number	Definition	MIL On?
13	Left (Bank 2) oxygen sensor circuit error	Yes
14	Engine coolant temp. sensor circuit, high temp.	Yes
15	Engine coolant temp. sensor circuit, low temp.	Yes
16	System voltage discharge (low battery voltage)	Yes
17	Camshaft position sensor error	No
21	Throttle position sensor circuit, high signal voltage	Yes
22	Throttle position sensor circuit, low signal voltage	Yes
23	Intake air temp. sensor circuit, low temp. indicated	Yes
24	Vehicle speed sensor circuit, no signal	Yes
25	Intake air temp. sensor circuit, high temp. indicated	Yes
33	Manifold absolute pressure sensor circuit, high signal voltage = low vacuum	Yes
34	Manifold absolute pressure sensor circuit, low signal voltage =high vacuum	Yes
35	Idle air control circuit error	Yes
36	24X signal circuit error	No
39	Clutch switch error	Yes
42	Ignition control error	Yes
43	Knock sensor error	Yes
44	Left (Bank 2) oxygen sensor, low voltage =lean exhaust indicated	Yes
45	Left (Bank 2) oxygen sensor, high voltage =rich exhaust indicated	Yes
46	Anti-theft system out of frequency range	No
51	PROM error, faulty or incorrect calibration	Yes
53	System voltage overcharge (high battery voltage)	Yes
54	Fuel pump circuit, low voltage	Yes
61	A/C system performance problem	No
63	Right (Bank 1) oxygen sensor circuit error	Yes
64	Right (Bank 1) oxygen sensor, low voltage = lean exhaust indicated	Yes
65	Right (Bank 1) oxygen sensor, high voltage = rich exhaust indicated	Yes
66	A/C refrigerant pressure sensor circuit, low pressure	No

Code Number	Definition	MIL On?
67	A/C refrigerant pressure sensor, circuit error	No
68	A/C relay circuit, shorted	No
69	A/C relay circuit, open	No
70	A/C refrigerant pressure sensor circuit, high pressure	No
71	A/C evaporator temp. sensor circuit, low temp.	No
73	A/C evaporator temp. sensor circuit, high temp.	No
75	Digital EGR valve circuit, # 1 solenoid error	Yes
76	Digital EGR valve circuit, # 2 solenoid error	Yes
77	Digital EGR valve circuit, # 3 solenoid error	Yes
81	Torque converter clutch brake switch circuit error	No
82	Ignition control 3X signal circuit error	No
85	PROM error, faulty or incorrect calibration	Yes
86	Analog/Digital translation error in ECM	Yes
87	EEPROM error	No

On this vehicle model, the Malfunction Indicator Lamp displays "Service Engine Soon."

Figure 1–3 Trouble Code Definitions: 1993 Chevrolet Camaro/Pontiac Firebird With 3.4L V-6 Engine.

The first OBD II vehicles were introduced in 1994, as the transition from OBD I began in earnest. Because of the wide array of differences in OBD I systems, 1994 and 1995 models were built with varying degrees of OBD II compatibility as automotive engineers were challenged to make their existing software and designs in other vehicle systems to work with the OBD II lines of communication, known as **protocols.**

The final OBD II mandates were enacted by Congress in 1997. All automotive manufacturers were required to provide the aftermarket with access to emission control related technical information, software reprogramming capability, all necessary test equipment, and training materials.

Benefits of OBD

With the introduction of OBD systems, the repair technician was armed with a new set of tools to provide guidance in diagnosing and repairing problems in electronic engine management and emission control systems. The vehicle was now capable of telling the technician which signals, components, or functions went out of their normal range of operation while the computer was "watching" the engine run. The engine control computer was called a variety of names during the OBD I years, including Electronic Control Unit (ECU), Multifunction Control Unit (MCU), Engine Control Module (ECM), Single Board Engine Controller (SBEC), Single Module Engine Controller (SMEC), and a system used by Chrysler Corporation that used a pair of microprocessors called the Logic Module and the Power Module. In the OBD II system, the computer is universally known as the **Powertrain Control Module (PCM).**

Comparison of OBD I to OBD II

In brief, OBD II is, as its name implies, a second generation on-board diagnostic system. It is an enhanced diagnostic monitor, with expanded capabilities and a different primary purpose than that of OBD I. As previously stated, some version of OBD I was present in all cars beginning in the 1988 model year; a universal OBD II system has been present in all cars since the 1996 model year.

While OBD I was primarily designed to detect and alert the driver to malfunctions that would cause driveability problems, OBD II is designed to detect virtually any problem that would cause an increase in exhaust emissions due to an electrical, chemical, or mechanical malfunction. Here is an example that illustrates this basic difference:

On an OBD I-equipped vehicle, let's assume that the catalytic converter stops working, but everything else is functioning correctly. Perhaps the catalyst has become contaminated from anti-freeze due to a blown head gasket, or it may have just become unable to function after 100,000 miles of use. In any case, all systems and components other than the catalytic converter are functioning as designed electronically, and the computer is making all the necessary adjustments. The engine is running in closed loop, so the MIL will not light, and the vehicle driver will not know there is a problem. However, the chemical cleaning of the exhaust by the catalytic converter is not happening, and the emissions at the tailpipe will be increased. OBD II will detect a scenario like this and alert the driver to the malfunction.

The OBD II system uses all of the diagnostic features of OBD I, then adds monitors to check the efficiency of the catalytic converter and to check for mechanical and electrical malfunctions that OBD I could not measure. A tailpipe emission increase could be caused by a problem that will not result in any detectable problem with the vehicle's driveability. The ability to detect and alert the driver to this type of malfunction is the major difference found in OBD II. So, the major concern of OBD I is driveabilty, while the major concern of OBD II is unacceptable emissions. OBD II is designed to report any malfunction that *may* cause exhaust emissions to increase to 1.5 times the levels the car was certified for by the EPA, or any malfunction that will cause the catalytic converter to be damaged.

As previously discussed, the history of vehicle emission regulations and the auto manufacturers' response to them was quite uneven and somewhat less than well-structured. Over the years, a number of exhaust emission testing and inspection programs have been started, stopped, redesigned, enhanced, and otherwise modified in a series of confusing changes that resulted in questionable outcomes. In most cases, emission inspection and repair legislation were enacted in an attempt to address the problem of *nonattainment areas*, local regions that did not meet the EPA's air quality standards when tested. Almost without exception, these regulations were extremely unpopular politically speaking!

OBD II has the advantage of placing the emissions tester right in the vehicle, as well as providing ongoing testing as the vehicle is driven. This is much more comprehensive than an annual drive-through emissions test required to renew a vehicle's registration. OBD II is so effective in detecting increased emissions that an OBD II "plug-and-play" test for stored trouble codes has replaced the traditional exhaust tailpipe emissions test in many parts of the United States. Anything from a misfiring spark plug to a loose gas cap will be detected and stored in the memory of the PCM by the OBD II system. The MIL is the interface that the PCM uses to communicate to the driver that something needs attention. With OBD II, the MIL reports the urgency of the malfunction by changing the way in which it displays the trouble. See Figure 1–4.

- **Single Quick Flash of the MIL:**
  A single quick flash of the MIL indicates a momentary malfunction. No action is necessary and the driver may not notice the lamp or the problem.

- **Steady Illumination of the MIL:**
  A MIL that remains illuminated indicates an ongoing problem that needs attention. The vehicle should be checked as soon as possible to determine and repair the cause.

- **Rapid Flashing of the MIL:**
  A rapidly flashing MIL indicates existing conditions that can severely damage the catalytic converter. The vehicle speed and load should be reduced immediately. If this does not revert the MIL to steady illumination, the engine should be shut down and the vehicle should be towed to a repair facility.

Figure 1–4 The MIL reports trouble three ways.

By law, the MIL *must* illuminate for any problem that *might* increase exhaust emissions by 50 percent or more above the levels measured in the Federal Test Procedure for the car when new. The MIL also *may* illuminate for problems that do not affect tailpipe emissions, at the discretion of the manufacturer. Most vehicles are designed with enhanced diagnostics that will turn the MIL on for a variety of faults not related to emissions.

While OBD I trouble codes were read by counting the "flash codes" on some models, OBD II requires the use of a scan tool or personal computer with OBD II software interface to read the trouble codes and data displays on all models beginning with the 1996 model year. The good news is that all vehicles with OBD II use the same size and shape of connector and provide the same minimum level of data and trouble codes to the repair technician. Virtually all auto manufacturers have expanded the capabilities of their OBD II systems to enable monitoring and diagnosis of chassis and body systems beyond the requirements of OBD II regulations.

The OBD II DLC connector also is *supposed* to be located in the same area on all vehicles, although some manufacturers have continued to insist on being creative in locating the DLC (Figure 1–5). The regulations require that either the DLC or a label directing the technician to the DLC must be located below the dashboard on the left-hand side of the vehicle. When seeking the DLC, it is important to look carefully, as some manufacturers will hide the connector behind a removable panel or other component, while others are readily visible under the dash. In a few cases, a label will direct you to another location, such as a center console trim panel or ashtray that must be removed to access the DLC.

Refer to Appendix D for a list of non-standard DLC locations used by various manufacturers. DLC locations are generally indicated within electrical component

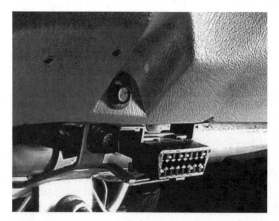

Figure 1–5   The DLC is not always where it is supposed to be. The 1999 Chevrolet Suburban uses a preferred location, but the 1998 Volvo S70 is more creative.

locator reference charts. In addition, many scan tool software programs show DLC locations when the vehicle is selected by year, make, and model, rather than using *Generic OBD II* when setting up the test routine.

The Evolution Continues: OBD III

In reviewing the technological advances that have occurred in automotive engine management systems to date, it would be very shortsighted to assume that we are anywhere near the peak of on-board, self-diagnostic capabilities. The next generation, OBD III is already past the drawing board, and already technologically possible as we go to press.

There are several major differences between OBD II and OBD III, but several similarities as well. All of the vehicle's data and trouble codes captured by OBD II are still available with OBD III, but new enhancements have been added, as well as new ways of communicating the data. The greatest difference is that, while OBD II requires a scan tool or similar device to be attached to the vehicle's DLC, OBD III can read all available vehicle data remotely, without the vehicle having to be present! Wireless transmission of vehicle data will make drive-through emissions testing a thing of the past. Where required, emissions testing can be accomplished by simply driving past a wireless "hot spot," similar to the electronic devices used on toll roads. Emergency road service providers will be able to determine what caused a vehicle to become disabled before sending a technician or tow truck to the scene. Data not directly related to emissions, such as the need for oil changes, maintenance indicators, and vehicle operating conditions, could be transmitted directly to repair facilities.

Other uses for the wireless interface devices of OBD III also are technologically possible, and some of them are controversial. For example, dialing into a vehicle's data transmission frequencies could allow law enforcement personnel to shut off the engine of a car they were pursuing. However, this same technology could be used for dark purposes by those with criminal intentions, essentially preventing their intended victims from escaping. Daily routines and driving conditions could be transmitted to transportation authorities to assist with traffic control and highway incident management, but this data also could be used to report traffic law violations and vehicle abuse.

All of this is being debated as we go to press. It is generally believed that OBD III will soon be the standard for new vehicles, but there is much to be resolved with respect to legal issues and security of the vehicle owner before it is rolled out. At this time, it appears that OBD III may make its first appearance on fleet vehicles, where time scheduling and profiling can make their operation most efficient.

In the next chapter, we will begin to examine the language of OBD II and learn more about how it operates and interrelates to the vehicle's engine and drive train.

Review and Reinforcement Questions

1. The first emission control device installed on vehicle sold in the United States was:

   a. Catalytic converter

   b. EGR valve

   c. EFE valve

   d. PCV valve

2. Reading of OBD I diagnostic trouble codes without the use of a scan tool was generally accomplished using:

   a. Digital multimeter

   b. Analog multimeter

   c. Flash codes

   d. Dummy codes

3. OBD I was primarily designed to detect driveability concerns, while OBD II is designed to detect problems causing excessive exhaust emissions. True or False?

   a. True

   b. False

4. OBD I would be likely to detect malfunctions in all of the following *except*:

   a. Oxygen sensor

   b. Manifold absolute pressure (MAP) sensor

   c. Engine coolant temperature (ECT) sensor

   d. Catalytic converter

5. What percentage of exhaust emissions increase is required by law to light the malfunction indicator lamp (MIL) on an OBD II vehicle?

   a. 10 percent

   b. 40 percent

   c. 50 percent

   d. 67 percent

6. A MIL flashes off and on repeatedly while accelerating and stays on steady at idle. This indicates:

    a. A severe misfire

    b. Likely damage to the catalyst

    c. Both of the above

    d. Neither of the above

2

OBD Terminology and Communication

Let's begin our exploration of the finer points of OBD II by taking a look at its language and basic functions.

Standardized Terms

One of the primary goals of the OBD II diagnostic system is standardization. A single, universal DLC size, shape, and terminal configuration allows a single scan tool connector to plug into any OBD II–compliant vehicle. Any given trouble code number has the same definition on every OBD II model. Terminology, acronyms, and abbreviations also are standardized with OBD II. This standardization process and its effects of streamlining and simplifying the many previous diagnostic hookups and routines is one of the most important features of the generic OBD II system. Refer to Appendix B for a list of standard OBD terms and acronyms as provided in the Society of Automotive Engineers (SAE) recommended practices for OBD II design in document J-1930.

OBD II Communication

Protocols

Another form of standardization lies in the self-diagnostic and monitoring features found in all OBD II–compliant vehicles. OBD II provides a veritable arsenal of information and data to aid the technician in troubleshooting and verifying correction of a large number of malfunctions. The OBD II system actively checks the operation of the engine's fuel delivery, ignition, emission controls, and basic mechanical systems on an ongoing basis.

As previously discussed, individual manufacturers are still allowed to provide information above and beyond the generic OBD II codes and data required by law. Virtually all manufacturers do provide additional information and features, but this information will not be readable by some scan tools. In addition, some of the additional content is allowed to be proprietary and may be used only by the manufacturer's own diagnostic equipment and software.

There are three different lines of communication commonly used by various manufacturers, each requiring a different software interface. These lines of communication are called **communication protocols**. A given vehicle will use only one protocol, but the universal 16-pin DLC and scan tool plug are designed to accommodate all of them. Two protocols were initially developed by SAE International and one by the International Standards Organization (ISO). Numbers associated with the protocols refer to the organization's documents from which the protocol standards were drawn.

The three current communication protocols and the DLC pins used to communicate data in each of them can be seen in Figure 2–1. Also shown are the uses of the remaining pins assigned to OBD II. You will notice that pins 1, 3, 8, 9, 11, 12, and 13 are not assigned. These pin locations may be used by the vehicle manufacturer for any desired purpose, such as proprietary reprogramming or diagnosis of multiple subsystems and accessories not related to OBD II.

Who Uses Which Protocol?

As we go to press, the use of OBD II protocols by manufacturer is as follows:

- ISO-9141-2 protocol was used by DaimlerChrysler domestic vehicles exclusively until 1998, when phase-in to J1850 VPM began. It is still used for some models by DaimlerChrysler, as well as most European and nearly all Asian vehicles. This protocol uses pins 4, 5, 7, 15, and 16.

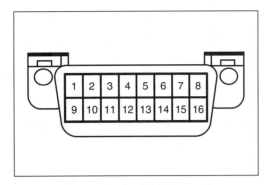

Figure 2–1 DLC Legend:

Pins 1,3,8,9,11,12,13	OEM Discretionary Use (Unassigned)
Pin 2	SAE J-1850 Bus Positive (+)
Pin 4	Chassis Ground
Pin 5	Signal Ground
Pin 6	CAN High (SAE J-2284)
Pin 7	ISO-9141-2 "K" Line
Pin 10	SAE J-1850 Bus Negative (−)
Pin 14	CAN Low (SAE J-2284)
Pin 15	ISO-9141-2 "L" Line
Pin 16	Vehicle Battery Power (+)

- SAE J-1850 PWM (Pulse Width Modulation) protocol is used by Ford Motor Company domestic vehicles. This protocol uses pins 2, 4, 5, 10, and 16.

- SAE J-1850 VPM (Variable Pulse Width Modulation) protocol is used by most General Motors vehicles and some DaimlerChrysler models since 1998. This protocol used pins 2, 4, 5, and 16.

- Manufacturers are in the process of phasing-in Controller Area Network (CAN) bus protocol. All vehicles sold in the 2008 model year must be equipped with this protocol.

Most professional-grade scan tools are designed to check for the correct protocol to read. This protocol is automatically selected as the diagnostic software is initialized. However, it is important to note that some of the less expensive scan tools do not have this capability, especially those sold for consumer use. In addition, some scan tools are only capable of reading and displaying diagnostic trouble codes and cannot display a data stream.

Another important variation in scan tool capabilities is the ability to provide repair and troubleshooting tips. You can guess which tools have these features—the more costly models!

Discretionary Information

As illustrated in Figure 2–1, the DLC pins numbered 1, 3, 8, 9, 11, 12, and 13 are not assigned to any OBD II functions. These pins may be used for any desired purpose by vehicle manufacturers, or not used at all. As you might guess, their use is not standardized. For example, a Chrysler vehicle may use pin 1 for keyless entry system enable, while a Ford vehicle may use pin 1 for ignition control. Chrysler may use pin 8 for switched battery power, while the Ford uses pin 9 for that purpose.

In general, some combination of the unassigned DLC pins will be used by the manufacturer to communicate with dealer service department level scanner equipment fitted with the manufacturer's software. This level of communication allows access to a large amount of information about the various computer-controlled systems found on the vehicle. Enhanced diagnostics and other functions provided may encompass everything from electronic suspension troubleshooting to programming the bass and treble ranges of the audio system.

Scan Tool Interface

The line of communication between the information stored in the PCM's memory and the repair technician is called an **interface**. With respect to troubleshooting and diagnosis, the scan tool is the link used to allow the technician to see what the OBD II system has available to view. Exactly what data can be examined and what tests can be run will vary considerably from one vehicle to another and one scan tool to another.

You may ask, "How is this possible? Isn't OBD II supposed to be standardized?" The answer is that OBD II is standardized at the minimum level of communication required for compliance. This minimum level of standardization applies to both scan tool design and PCM diagnostic programming. So, many manufacturers include substantially more information than required in their OBD II diagnostic systems. Likewise, the capabilities contained in scan tool software are enhanced to take advantage of the additional data and functions allowed.

OEM versus OBD II Interface

Nearly all professional-grade scan tools are capable of running two types of diagnostic programs: the generic OBD II trouble codes and data stream required by law, and the codes and data stream program designed by the vehicle manufacturer (OEM). Since the manufacturer's program contains proprietary information and features, there will be limits to what the scan tool can access and the limits will

OBD II Generic	OBD II Dealer Service Department	OEM Proprietary Tests (Non-OBD II Interface)
Read powertrain DTCs	Increased data parameters	Data stream
Read Freeze Frame data	Manufacturer—specific DTCs and descriptions	Actuator Output Tests
Erase DTCs		Trouble codes (OBD I)
Erase Freeze Frame	Status of monitors	
Read Readiness Status	Status of DTCs	
Oxygen Sensor Monitor	Similar conditions window	
Continuous Monitors*	Reprogramming capabilities†	
Noncontinuous Monitors*	Diagnosis and data for multiple systems and accessories other than OBD II	
Powertrain data stream		
Parameter ID display*	Interface with Bus Communications Network	

*Available information in these categories vary from one year, make, and model to another, and from one scan tool to another.

†Limited reprogramming capabilities are now available to the aftermarket, but not directly through OBD II. Removal of PCM from vehicle is generally required.

Figure 2–2 OBD II vs. OEM Scan Tool Function Comparison.

vary somewhat from one make and model to another. With most scan tools, the greatest difference is that no troubleshooting and repair tips can be accessed with the generic OBD II program. Figure 2–2 shows a comparison of the information and functions typically available with a professional repair industry scan tool and a proprietary diagnostic system, such as those found in a dealership's service department. You will note that the OEM software interface must be used to test actuators (output tests) and systems not related to OBD II, such as the automatic transmission and antilock brake system (ABS).

There also is a difference between the OEM and OBD II software interfaces in the amount of detailed data categories that can be accessed. The abilities that go beyond the generic OBD II level are referred to as *enhanced diagnostics*.

Whether you are using the OEM proprietary system or the generic OBD II information in your diagnosis and troubleshooting, you will still need access to specific repair information and data to know what the readings mean and what you should do with them. Sources include manufacturers' repair manuals, aftermarket databases, and diagnostic/troubleshooting software that may be added onto a scan tool, lab scope, or personal computer. It is important to remember that the trouble codes displayed are not the end of the diagnostic process, but rather the beginning!

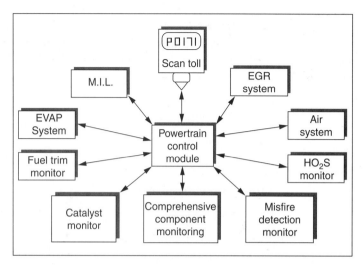

Figure 2–3   PCM Inputs and Outputs.

Basic Engine Management Systems

The basic operation of computer-controlled engine management systems is quite similar for all vehicles, regardless of the specific design differences from one year, make, and model to another. All computer-controlled vehicles, dating all the way back to the late 1960s, function the same way in that the computer monitors various input signals and uses these values to decide the proper outputs to apply to the systems and components it is responsible for controlling. Figure 2–3 illustrates how a typical engine control computer functions in the operation of engine and emission control components.

PCM Operation

Now, let's look at the operation of the engine management system and the PCM, in particular. It is important to understand how the PCM functions and the interrelationships that exist between the PCM, the various inputs it relies on in decision-making and the outputs it produces.

Types of Memory

Like any computer processor, the PCM's capability to make decisions concerning fuel delivery, spark advance, and operation of emission control components is controlled by sets of instructions programmed into it by the manufacturer. There are three basic types of memory contained in the PCM: Read-Only Memory (ROM), Random Access Memory (RAM), and Keep-Alive Memory (KAM).

The ROM may be thought of as a library of reference materials. It contains the operating instructions to tell the PCM what to do when the information it receives consists of a given combination of signals. Based on the inputs received, the PCM will react to control the outputs that it has control of. The decisions may include such items as:

- How far to open the EGR valve
- Whether to provide additional fuel or less fuel than the basic, preprogrammed amount of fuel delivery
- Whether to inject air into the exhaust system upstream, at the exhaust manifold, or downstream, into the catalytic converter
- How far to advance the ignition spark timing (sometimes, each cylinder is timed individually)
- Where to position the **idle air control valve**, or other idle speed control device
- When to shift gears of the automatic transmission

Numerous other component operations may be controlled by the PCM, depending on the specific year, make, and model of the vehicle and the exact type of components with which it is equipped. Many vehicles are equipped with sections of ROM that can be reprogrammed or customized to a particular vehicle with a specific engine and equipped with specific options. This type of ROM is referred to as Programmable Read-Only Memory (PROM.). Variations of PROM include Electronic Programmable Read-Only Memory (EPROM) and Electronic Erasable Programmable Read-Only Memory (EEPROM). Another commonly used name for programmable memory is Flash PROM. All of these are basically the same type of memory, that is, reference instructions for the PCM that can be changed, updated, or customized. This feature allows a single PCM design to be used on numerous different vehicles and allows the PCM's programs to be updated by the manufacturer without having to replace the ECM. Some years ago, reprogramming the PCM could only be done using the vehicle manufacturer's proprietary software. However, software to reprogram PCMs and to correctly program a replacement PCM for a particular vehicle is now available in the aftermarket.

The RAM is the PCM's "note pad." Incoming information, much of which is ever-changing while the vehicle is being driven, is captured and relayed to the PCM. The PCM reads the RAM information and decides what it needs to add, subtract, move, advance, close, open, etc., by referring to the operating instructions contained in the ROM. Every time the information captured by the RAM changes, the old "notes" are erased and replaced with the new values.

The KAM is, as its name implies, responsible for "keep alive" functions. This section of PCM memory is actually a subsection of RAM that is independently powered directly from the vehicle battery. The KAM keeps key RAM information from being erased when the engine is turned off. For example, Long-Term Fuel Trim is stored, but Short-Term Fuel Trim is not.

Inputs and Outputs

Information that the PCM receives and uses in controlling the functions it is responsible for are called **inputs**. The commands it issues to control these functions are called **outputs.**

Inputs come in a variety of forms, due to the variety of engine conditions that must be accurately measured. Several different types of sensors are used to provide input information, and that information must be translated by the PCM into a common format so it can process and react to everything it needs. Typical sensor types used to communicate inputs to the PCM include:

- Thermistors, used to measure temperature
- Potentiometers, used to measure physical movement
- Transducers, used to convert pressure readings to electronic values
- Hall Effect switches, used to determine the exact position of moving components, such as the crankshaft
- Optical readers, used for the same purpose as Hall Effect switches
- Piezoelectric signal generators, used to create electrical signals when specific mechanical conditions are present, such as *detonation (knock sensor)*
- Magnetic signal generators, used to measure the speed of a moving component, where exact position is unimportant; commonly used for the Vehicle Speed Sensor (VSS) and ABS wheel speed sensor

Now, let's look at how these inputs work electronically. Although there are exceptions, two basic types of circuits are used with most sensors. One type uses a two-wire circuit that is connected directly to the PCM. This type of circuit is typically used for the sensors used to generate an alternating current voltage, such as the knock sensor, VSS, and ABS wheel speed sensor, and for single-resistor sensors such as the Engine Coolant Temperature (ECT) and Intake Air Temperature (IAT).

The second basic sensor circuit uses a three-wire connection. One wire is grounded, one is supplied with a "reference voltage," usually 5 volts, and the third is the "signal" communicated to the PCM. These sensors function by gradually reducing the reference voltage as a mechanical movement takes place within the sensor. For example, the Throttle Position Sensor (TPS) signals voltage changes as the throttle is moved, either toward wide open throttle or toward released (idle) position. This type of circuit is commonly used for TPS,

MAP sensors, and Hall-Effect sensors for functions such as crankshaft position. The three-wire circuit also is used for sensors which are variable resistors, such as the ECT and IAT sensors.

In general, the reference voltage, signal return, and ground are all connected to the PCM. Therefore, it has access to the entire sensor circuit and can often tell the difference between a circuit problem and a sensor malfunction. An additional feature used on some sensors, is **bias voltage**, which is a steady, regulated voltage applied to the signal wire by the PCM. When a bias voltage is used, the PCM uses it as a starting point for its calculations concerning that sensor. The PCM measures how much voltage is added or removed from the bias to determine the signal reading. In the case of most Oxygen Sensors, the bias voltage is 0.450 volts, the exact point of change from rich to lean air/fuel mixture, or 4.75 volts. It should be noted that normal bias voltage is not universal. It depends on sensor or system design on a given vehicle. However, it is usually between 2 and 5 volts.

Typical input data received by the PCM include:

- Engine RPM
- Engine Load (MAP or MAF)
- TPS
- IAT
- ECT
- VSS
- Intake Airflow (Airflow Meter or MAF)
- Brake on/off
- A/C request on/off
- A/C compressor on/off
- Short-term fuel trim
- Long-term fuel trim
- Engine knock (detonation)
- Misfire
- Crankshaft position
- Camshaft position
- Upstream Oxygen Sensor(s), (HO_2S1)
- Downstream Oxygen Sensor(s), (HO_2S2)
- Transmission input RPM
- Transmission output RPM

Outputs

The outputs of the PCM are the commands it generates in operating all of the components it controls. These components are adjusted, switched on or off, moved, or energized at the appropriate moments as directed by the PCM programming.

Typical PCM powertrain control outputs include:

- Fuel injectors "open" time (pulse width)
- EGR valve position
- Air injection amount and location
- Idle speed control
- Cruise control throttle operation
- EVAP system canister purge
- EVAP system pressure apply and release
- Fuel pump operation
- Ignition coil operation
- Ignition spark advance
- Automatic transmission solenoids (gear selection/shift points)
- Automatic transmission torque converter clutch (TCC)

In addition to engine management functions, the PCM may be responsible for a wide variety of outputs concerning the operation of heating and air conditioning, audio systems, power mirror and seat memory, electric door locks, antitheft system, and other accessories. In some cases, the PCM interacts and shares information with other control modules, which have specific system responsibilities. Examples include the Body Control Module (BCM), Antitheft Control Module, Antilock Brake System/Traction Control Module, and Electronic Suspension Control Module.

Outputs Are Also Inputs

When the PCM issues a command, the expected change in engine operation caused by that command is taken into consideration to adjust other outputs. For example, when the EGR valve is opened, exhaust gases enter the combustion chamber and displace some of the oxygen. In turn, the Oxygen Sensors will sense less oxygen in the exhaust stream, so the PCM compensates for the change in combustion by decreasing fuel delivery and perhaps, adjusting the spark advance. Depending on the engine operating conditions and how far the EGR is opened, intake manifold vacuum may decrease, a change that may be seen by the PCM as increased engine load. In turn, the PCM would then increase fuel delivery. When the air conditioning compressor is turned on, the engine load is increased. The PCM may add a bit more fuel delivery and will increase the engine idle speed to compensate. In effect, the output command also serves as input information. The PCM's reaction to output changes is an indication of the many interrelationships that exist between the systems and components on today's vehicles.

An important feature of PCM operation that also relates to the OBD II system is its capability to assess whether or not an output command actually attained the expected result. This is one way in which the functions under the PCM's control are monitored. For example, if the PCM increases the opening of the EGR valve, it looks for a change to indicate that the command was successful. This change may be measured by monitoring oxygen sensor signals, change in exhaust system pressure, or use of a position sensor mounted directly on the EGR valve assembly. Many earlier computer-controlled systems did not have this capability, especially those built before the days of OBD I.

In all OBD II–compliant vehicles, the PCM is designed to detect problems in its monitored functions that would affect driveability or exhaust emissions, although the law only requires the MIL to be illuminated for emission-related failures. It is extremely important to remember that OBD II is a series of diagnostic programs built into the PCM. It is *not* the engine management system.

Two operating modes are available for controlling the operation of fuel delivery, ignition timing, and emission control components: Open-Loop and Closed-Loop.

In open-loop operating mode, the PCM ignores many of the input signals received from the various sensors and switches and functioned using preprogrammed, set values for the amount of fuel to deliver, the spark advance, and the position of various emission control valves, solenoids, and switches. Open-loop operation is the mode used during warm-up or immediately after start-up, when the Oxygen Sensors are not warm enough to provide accurate readings, and the catalyst is not fully functioning. The PCM also will revert to open-loop operation if it doubts the readings of the key inputs it needs to operate its outputs correctly. In most cases, the only inputs the PCM uses in open loop are coolant temperature, air temperature, atmospheric pressure, engine speed, engine load, and throttle position. If any of these inputs are substantially incorrect, the engine will likely run very poorly, if at all. See Figure 2–4.

In closed-loop operation, the computer takes all inputs into consideration in calculating its outputs. Essentially, the computer reacts to the readings of all of its input signals by adjusting the amount of fuel delivered, the ignition timing, and the position of the various emission control devices accordingly. It will do this *only* if all of the key inputs are within normal range. If any important input is seen as abnormal or unreliable by the computer, it will revert to open-loop operation. See Figure 2–5.

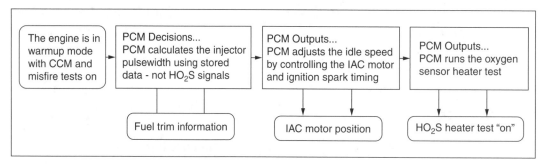

Figure 2–4   Operating Modes—Open Loop.

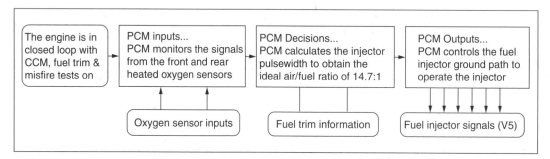

Figure 2–5   Operating Modes—Closed Loop.

It is important to realize that closed-loop operation still relies on preprogrammed values as "normal" settings. Then, the PCM adjusts the normal program for optimum operation based on the inputs. Although driveability symptoms may or may not be present in either open loop or closed loop, differences can be seen by watching the engine data stream on the scan tool. Such things as fuel trim and spark advance will almost certainly act at least somewhat differently when the PCM reverts to open-loop operation. Later on, we will discuss the details of open-loop and closed-loop operation as they relate to diagnostic routines.

Review and Reinforcement Questions

1. The lines of communication used by the PCM to exchange information with a technician interface, such as a scan tool are called:

    a. Bus lines

    b. Variable T-pulses

    c. Protocols

    d. CCD 1941 data streams

2. The abbreviations for the three primary types of memory used in the PCM are:

   a. RAM, PROM, and E-Bus

   b. RAM, ROM, and KAM

   c. RAM, EEPROM, and ROM

   d. ROM, PROM, and CAN

3. Which type of memory is used by the PCM as a "reference library" to tell it what to do with the inputs it receives?

   a. RAM

   b. ROM

   c. FCABM

   d. FTPM

4. The actions taken by the PCM in response to the information it receives are called:

   a. Inputs

   b. Outputs

   c. Control commands

   d. Response commands

5. The PCM relies on the oxygen sensor signal in which mode of operation?

   a. Warm-up

   b. Open loop

   c. Highway mode

   d. Closed loop

6. With OBD II, the PCM issues a command and checks to see if it was carried out properly. True or False?

   a. True

   b. False

3

OBD II System Monitors

Monitors are testing exercises, performed by the PCM, which are very carefully designed to indicate that all of the sensors within a portion of the engine management system are working properly to minimize emissions. In fact, OBD II monitors every component that is responsible for controlling emissions, directly or indirectly. For example, if the automatic transmission torque converter clutch fails to engage when commanded, exhaust emissions are affected. The PCM with OBD II will place the transmission into a limp-in mode and alert the driver by illuminating the MIL. To provide maximum efficiency, OBD II monitors divide the vehicle's engine management system into nine subsystems. Each subsystem has its own monitor. Each monitor is run by the PCM under very specific vehicle operating conditions. Because of this, all monitors may not run each time the vehicle is driven. A given monitor may not run for quite some time, until the exact combination of conditions occurs to trigger it. This set of conditions is called the **enabling criteria** for the monitor. With OBD II, the PCM has the ability to monitor multiple components and make accurate judgments about system efficiency. As we get deeper into troubleshooting tips and techniques later in the book, you will discover many useful monitors that you can perform yourself, with and without your trusty scan tool.

In an OBD II system, the following monitors are run by the PCM:

- Air Conditioning*
- Catalyst Efficiency
- Comprehensive Component Monitor (CCM)
- EGR System
- Evaporative System (EVAP)
- Fuel System (adaptive fuel system trim)
- Heated Catalyst
- Heated Oxygen Sensor
- Misfire Detection
- Positive Crankcase Ventilation*
- Secondary Air Injection
- Thermostat*

With the exception of Air Conditioning, Positive Crankcase Ventilation, and Thermostat, all are known as **major monitors.** We will look at each of the major monitors in detail later in this chapter.

Monitor Classifications

There are two broad classes of monitors used within the OBD II system, continuous and noncontinuous. Continuous monitors are always running when their enabling criteria are met and the vehicle is being driven. Noncontinuous monitors are only run when their individual enabling criteria are met, and the PCM does not detect any conditions or malfunctions that would cause the test results to be inaccurate. The enabling criteria for continuous monitors are much less detailed than those for the noncontinuous class. This difference exists because there are two levels of monitor priority. The most important job for OBD II to accomplish is to protect the catalytic converter from damage so it will continue to function efficiently. The two primary causes of catalyst failure are too much fuel delivery and ignition misfire. Therefore, the continuous monitors are the Misfire, Fuel System, and Comprehensive Component monitors. The remaining monitors are not quite so critical, so they are run only when all conditions are exactly right.

---

*These monitors are recent additions, not found on earlier OBD II vehicles. The Air Conditioning monitor is required by the State of California only if a vehicle uses R-12 refrigerant (Freon), but is used on a number of late model vehicles as well.

The Task Manager

The OBD II PCM software must manage a large amount of information and control a large number of functions. Built into the software is a program that acts a bit like a traffic controller. This feature is called different names by different manufacturers. Perhaps the most appropriate name is the one chosen by Chrysler, *Task Manager*, because that is exactly what it does with respect to the nine OBD II monitors. Deciding if and when each monitor is run is one of the jobs of the task manager. In addition, it is assigned to:

- Manage the operation of the monitors to prevent them from interfering with engine management and driveability
- Assure that the monitors are run in the correct order
- Prevent conflicts between any of the monitors
- Provide communication to the scan tool

Monitor Operating Modes

As the task manager is establishing its test priorities and deciding which monitors to run, there are four modes of operation where it places them—*Normal, Pending, Conflict, and Suspend.*

All of the modes except Normal will cause one or more of the monitors to be postponed.

- The Pending mode occurs if a sensor or signal that is malfunctioning is needed to run the monitor. The test will be postponed "pending" the repair of the problem.
- The Conflict mode is just what its name implies. The task manager sees that a conflict will occur if it allows two monitors it wants to run at the same time. It will postpone one of the monitors until the other one is completed in order to prevent inaccurate information from being captured.
- Suspend mode is simply a programmed delay to assure that the monitors are run in the proper order and under the right basic operating conditions. For example, the PCM needs accurate signals from the oxygen sensors in order to perform the Catalyst Efficiency Monitor. So, the Catalyst Monitor will be "suspended" until after the Oxygen Sensor Monitor has been run and satisfactorily passed, along with any other monitors that are required to provide information for this one.

Trips and Warm-up Cycles

In general, the task manager begins by looking for the necessary components of a *trip* before running any of the monitors. A trip exists when the vehicle is started and driven so that the enabling criteria are met for the monitor it wants to run. When this has happened and the monitor is completed, then the key is shutoff, it counts as one trip with respect to that particular monitor. The combination of driving conditions that enable the monitor and allow it to be completed is called a **drive cycle.** A drive cycle that allows enabling and completion of all of the monitors is called a **global trip.** The trip counter is used for a number of functions performed by the PCM including the operation of the MIL.

Trips are important for several reasons:

- Some catalyst-damaging faults, such as a severe misfire or electrical signal failure, will turn on the MIL in a single trip as soon as the fault is detected.

- Some faults require the same failure on two consecutive trips before the MIL is turned on, but a diagnostic trouble code is stored in memory after the first failure. The two-trip faults are those that are likely to cause damage, and more likely to be intermittent in nature.

- In general, any failure that turns on the MIL and stores a DTC will automatically capture **Freeze Frame data,** a snapshot of all of the PCM's data stream readings at the instant when the fault was first detected. This data can be displayed on the scan tool to aid in troubleshooting.

- If a high priority fault is detected, the PCM will look for the fault to occur within similar load, speed, and temperature conditions. This feature is called a *similar conditions window.* If the fault is detected again within the next 80 trips, a trouble code will be stored, although the MIL may not be turned on.

- Generally, at least three "good" trips, free of malfunctions, are required for the PCM to turn off the MIL once it has been turned on. Some stored trouble codes will require more than three trips before the MIL will be turned off.

- Any diagnostic trouble codes will eventually be erased from the PCM memory. This feature is important because it is designed to eliminate unnecessary troubleshooting of codes that were stored weeks, or even months before. The length of time a given code will remain stored varies from one trouble code to another, but at least 40 *warm-up cycles* (the part of a trip including engine start-up and reaching of normal operating temperature) are required. The PCM keeps track of warm-ups as well as completed trips and global trips. To count as a warm-up cycle, the engine must be run long enough to increase its temperature by at least 40°F *and* the temperature must cross the threshold of 160°F.

The ABCs of Monitors

Now, let's look at the individual major monitors in detail. We will examine the continuous monitors first and then move into the noncontinuous class. Keep in mind that the enabling criteria and type of sensor used will be similar for all manufacturers, but not identical. Manufacturers are still free to use their own designs and software programming, as long as the OBD II standards are met. Some vehicles require a very specific combination of temperature, speed, and load changes to be included in a drive cycle in order to trigger a particular monitor. See Figure 3–1. To make things more interesting, the drive cycle changes from one monitor to another and one vehicle to another. This is one more reason that good service information is absolutely essential to perform detailed OBD II diagnosis and repair. We will show some examples of specific drive cycles a little later when we discuss diagnostic routines. The following operating details and sensors identified are typical.

The first three monitors we will examine are continuous monitors.

The Misfire Monitor

The Misfire Monitor is continuous, but remember that there are still enabling criteria that must be met. However, there are fewer criteria for the Misfire Monitor than for any other, and also fewer reasons for it to be postponed. This is true because a severe misfire can damage the catalytic converter very quickly. Raw fuel entering the catalytic converter will cause serious overheating and literally burn the catalyst into useless, black lumps. In addition, less severe misfire will increase unburned hydrocarbon emissions.

The enabling criteria for the Misfire Monitor are:

- Valid signals from MAP, MAF, ECT, VSS, and RPM
- RPM within a specified range
- ECT within a specified range (generally does not require a full warm-up cycle)
- VSS within a specified range

Note   Closed-loop operation is *not* required to allow the Misfire Monitor to run in most vehicles.

The following conditions will cause the Misfire Monitor to be postponed:

Pending

- Waiting for a critical input signal that is not present, including RPM, VSS, MAP, TPS, CMP, or CKP. (A trouble code will likely be stored for the offending sensor.)

The following is an example of a typical General Motors drive cycle. All drive cycles are not the same, even within the same make and model year. Always look up the drive cycle for the exact vehicle being serviced and follow it to the letter!

A complete driving cycle should cause monitors to run on all systems. This particular drive cycle can be completed in 15 minutes.

Drive Cycle Routine:

1. **Cold start**   Engine temperature must be below 122°F (50°C) *and* within 11°F (6°C) of the ambient air temperature at start-up. Do not leave the ignition turned On prior to the cold start or the HO$_2$S monitor may not run.

2. **Idle**   Engine must be run for 3 minutes with the air conditioning and rear defroster turned On (if so equipped). In all cases, operate as many electrical accessories and systems as possible. This will test the HO$_2$S heaters, Secondary Air Injection, EVAP Purge "No Flow," Misfire and, if closed loop is achieved, Fuel Trim.

3. **Accelerate**   Turn all electrical accessories Off. Accelerate to 55 mph (88 km/hr) at half-throttle. During this step, Fuel Trim, Misfire, and Purge Flow will be tested.

4. **Hold steady cruise speed**   Drive at a steady speed of 55 mph (88 km/hr) for 3 minutes. During this step, HO$_2$S Response, Secondary Air Injection, EGR, Misfire, Purge, and Fuel Trim will run.

5. **Decelerate**   Release the throttle without depressing the brake or clutch. Do not shift gears. Coast down to 20 mph (32 km/hr). This step will test EGR, Fuel Trim, and EVAP Purge.

6. **Accelerate**   Increase speed to 55–60 mph (88–96 km/hr). Fuel Trim, Misfire, and Purge Flow will be tested.

7. **Hold steady cruise speed**   Drive at a steady speed of 55 mph (88 km/hr) for 5 minutes. During this step, Catalyst, HO$_2$S Response, Secondary Air Injection, EGR, Misfire, Purge, and Fuel Trim will run. NOTE: If the catalyst function is marginal or the battery has been disconnected, five complete drive cycles may be required to determine the state of the catalyst.

8. **Decelerate**   Release the throttle without depressing the brake or clutch. Do not shift gears. Coast down to 20 mph (32 km/hr). This step will repeat the EGR, Fuel Trim, and EVAP Purge monitors.

Figure 3–1   OBD II Drive Cycle Example.

- Engine management system is operating in limp-home mode, such as shutting off fuel and ignition to some cylinders in an attempt to reduce overheating.

Conflict

- Fuel system too rich or too lean, as indicated by oxygen sensor reading(s)
- EVAP purge was attempted and failed
- EGR system failure detected

Suspend

- Engine cranking but not yet running
- Cold start conditions
- Deceleration with fuel cutoff activated
- RPM out of test range
- Rapid "pumping" or throttle between open and closed
- MAP or MAF signal fluctuates rapidly
- Rough road detected (uses VSS and/or ABS wheel speed sensors)
- Level of fuel in tank too low

A misfire is measured using the engine RPM sensor, which usually takes its reading from the engine flywheel. The PCM looks for momentary changes in the RPM that are timed with the firing of the spark plugs. If the engine speed slows slightly at the exact time a spark plug should fire, it is seen as a misfire. The list of possible causes for a misfire condition is a long one and includes malfunctions in the ignition, fuel, and emission control systems, as well as basic mechanical problems. Common causes for a misfire DTC are:

- Worn, loose, or fouled spark plugs
- Defective spark plug wires
- Defective ignition coil
- Contaminated fuel
- Incorrect fuel pressure
- Disconnected, leaking or clogged fuel injector(s)
- Improper fuel control caused by faulty input from sensor, usually MAP, MAF, or ECT
- Defective CKP sensor
- Incorrect EGR flow
- Improper valve timing (slipped timing chain or belt)

- Worn camshaft lobes
- Sticking or burned valve(s)
- Improper valve clearance adjustment
- Sticking or worn valve mechanism (lifters, rocker arms, pushrods)
- Weak or broken valve spring(s)
- Loss of compression (pistons, rings, head gasket, cracked cylinder head or block)

It is important to realize that a "false misfire" may be caused by an extraneous condition such as a loose drive belt that slips intermittently or driving on a "washboard" road, especially if the vehicle has play in the driveline. Anything that can cause a fluctuation in the crankshaft speed may be perceived as a misfire if it occurs at the time the PCM expects a spark plug to fire.

DTCs and Action of the MIL

- DTC is stored when misfire is first detected that will increase emissions. MIL is turned on if misfire is severe. If not severe, MIL is not turned on after first trip.
- On second consecutive trip with nonsevere misfire, MIL is turned on and freeze frame data is captured.
- If severe misfire occurs (15 percent or more of 200 firings), MIL flashes on and off to signal likely catalyst damage. MIL will remain on if amount of misfiring reduces.
- MIL will turn off after three good trips, but only if they are in very similar driving conditions to those when misfire was detected.

Fuel System Monitor

The Fuel System Monitor is continuous, but requires more enabling criteria than the Misfire Monitor. This monitor is used to check the operation of the PCM's fuel correction program, which is a feature designed to add or take away fuel delivery from the standard PCM program to maintain the most efficient combustion. Various names are given to this feature by different manufacturers, but we will stick with the standard OBD II terminology.

Two forms of fuel correction programming are used. They are *Short-Term Fuel Trim* and *Long-Term Fuel Trim.* The two trim adjustments work together to adjust the amount of fuel delivered to the cylinders for optimum combustion. In so doing, they allow the engine to "learn" a new fuel delivery program custom-designed just for it. This is how they work:

First, the PCM divides the engine's operating conditions into "windows." Think of the windows as a row of files, where information is stored. Each window is assigned to capture information in a specific load and speed range of the engine. So, wide open throttle acceleration has a window, deceleration at closed throttle has another, cruising at 40 mph has another, part-throttle acceleration has another, and so on. As the driving conditions change, the PCM keeps track of which window the engine is running in, *while watching the Oxygen Sensors the whole time!* If the air/fuel mixture is too lean, the PCM adds more fuel to the delivery program and files that information in the appropriate window. If the mixture is too rich, it takes away some fuel and files the information. The process goes on for as long as the engine is running in closed loop. This is a simplified description of short-term fuel trim.

Now, while the short-term fuel trim is making adjustments and filing its reports, the long-term fuel trim program is watching the short-term activity. As it sees fuel delivery changing, it makes smaller, more permanent changes in the same direction as the short-term corrections, either richer or leaner. Although the short-term fuel trim is constantly changing, the long-term trim information is stored in the PCM's memory and applied to the basic fuel program.

Since these two fuel trim elements work together, the PCM uses the combined correction of both, or *total fuel correction*, to adjust the fuel delivery as needed. The end result is the ideal fuel delivery for keeping the engine in closed loop and providing the most efficient combustion.

The Fuel System Monitor functions by carefully comparing the corrected fuel delivery to oxygen sensor responses. Since engine speed and load, EGR position, and EVAP controls operation change frequently, the PCM has a number of different test modes in which to run this monitor. By now, you may have figured out that this is one very complex monitor!

The enabling criteria for the Fuel System Monitor are:

- Warm engine (above 160°F)
- Engine Management System operating in Closed-Loop mode
- Valid signals present from MAP, MAF, BARO, ECT, IAT, TPS, VSS, and RPM
- Long-Term Fuel Trim data present
- Short-Term Fuel Trim data present

The following conditions will cause the Fuel System Monitor to be postponed:

Pending

- MIL is on because of failure in any of the enabling criteria above or any of the following:
    - DTC stored for an upstream oxygen sensor or its heater
    - DTC stored for EVAP monitor or solenoid failure
    - DTC stored for EGR monitor or solenoid failure

Conflict

- DTC for misfire with no MIL
- DTC for EVAP monitor with no MIL
- DTC for upstream oxygen sensor heater with no MIL
- DTC for EGR monitor failure with no MIL

Suspend

- Fuel level in tank is too low.
- Any engine speed and load, vehicle speed, or EVAP system operation conditions that could affect the monitor results.

DTCs and Action of the MIL

- DTC will be stored if the monitor detects the fuel system running too lean or too rich, no MIL on first trip unless malfunction is likely to cause damage to catalyst. If so, MIL is turned on immediately.
- On second consecutive trip with rich/lean limits exceeded, MIL is turned on, DTC is stored and Freeze Frame data is captured.
- After three good trips in very similar operating conditions to those present when the failure occurred, the MIL may be turned off.

Comprehensive Component Monitor

In general terms, the Comprehensive Component Monitor (CCM) is used to check for short circuits, open circuits, and *rationality* of readings between the many inputs that the PCM relies on to manage the engine systems correctly, and a few outputs as well. In many respects, it is similar to the fault-detection monitors that were present in OBD I systems. You will recall that OBD I pretty much just watched and waited for the value of one of the signals to go out of its predetermined normal range. The differences in the OBD II version are that it runs continuously and that it does not include components such as the Oxygen Sensor and EGR, which have their own monitors to test them directly.

In the CCM, the components are evaluated in several ways. Sensor values and input circuits are monitored to detect short circuits and open circuits like the OBD I system did. In addition, the sensors will be tested to see how long they take to meet the enabling criteria after start-up. Finally, the readings of sensors will be checked to see if any of them do not make sense. For example, let's say that the outside air temperature is 80°F, and the engine has been running for 10 minutes. If the IAT shows 98°F but the ECT shows 75°F, the readings are in conflict with each other. Not only is it irrational for the coolant temperature to be lower than that of the intake air, it does not make sense for the coolant to be so cool after running for 10 minutes. This scenario would cause the ECT sensor to fail the rationality evaluation. The CCM is very good at "seeing the big picture" and picking out which input is irrational. This monitor is really quite similar to your own thinking as you look over a list of data on your scan tool display to find the "bad apple."

The enabling criteria for the Comprehensive Component Monitor are a bit vague, as well as being different for various inputs. They operate this way:

- Some components are tested immediately after "key On." This will happen even if the key is moved directly from "Off" to "Start." For example, the PCM will capture base-line temperatures and the BARO reading, and the CCM will be watching those inputs!

- Most other component tests require reaching closed-loop operation so the inputs and the PCM's response can be compared, although all inputs and some output devices will be checked continuously for shorted or open circuits, even in open loop. The exact criteria for monitoring the individual components vary substantially among manufacturers. This is one area where the engineers did not consult with each other!

- There are no other conditions that will cause the CCM to be postponed.

Inputs typically monitored by the CCM include:

- Brake On/Off Switch
- CMP
- CKP
- RPM
- Cruise Control Commands/Requests
- ECT
- IAT
- EVAP Switches
- Knock Sensor
- MAF
- MAP
- BARO
- Clutch Switch (manual transmission)
- Transmission Range Selector (automatic transmission)
- VSS

PCM outputs typically monitored by the CCM include:

- EVAP Purge and Vent Solenoids
- IAC or ISC position
- Ignition Spark Advance Controls
- Torque Converter Clutch Solenoid
- Automatic Transmission Shift Solenoids

The remaining monitors are of the noncontinuous type.

Oxygen Sensor and Oxygen Sensor Heater Monitors (HO$_2$S)

The oxygen sensor readings are the *most important information* that the PCM relies on in operating the engine management system and controlling fuel, ignition, accessories, and emission controls. From watching the action of each oxygen sensor, the PCM can "see" whether or not the exhaust emissions are being properly controlled, whether the catalytic converter is doing its job, and whether the engine is running efficiently. The oxygen sensor signals also are relied upon to run other monitors, including the Fuel System, Catalyst, EGR, and EVAP. In addition to the monitors that test their effectiveness, the oxygen sensors are checked continuously for shorted or open circuits.

Because the HO$_2$S signals are so important, it is critical that they start functioning as soon as possible. That is the purpose of the heater installed on all OBD II oxygen sensors, and the reason that the heater circuit has its own monitor.

The following elements of HO$_2$S operation are tested during the course of the monitors:

- Operation of the heater (This portion of the testing may be done after the engine is shutoff, while the exhaust temperature is cooling.)
- Acceptable signal voltage range when oxygen level changes
  - Low oxygen content = high signal voltage
  - High oxygen content = low signal voltage
  - Voltage level of 0.450 millivolt is "threshold" of lean versus rich
- How quickly the signal voltage rises and falls in response to oxygen content changes
- Evidence of an open or shorted circuit in the HO$_2$S circuit (voltage "sticks" in one range: high (shorted), low (open))
- Evidence of a "dead" HO$_2$S (signal voltage moves only slightly as oxygen content changes)

- Cross counts (how many times the signal voltage crosses the 0.450-millivolt threshold)

The enabling criteria for the Oxygen Sensor Monitor are:

- Engine temperature warm (usually above 120°F.)
- Engine run time exceeds preprogrammed limit (required time varies with IAT and ECT temperatures)
- Transmission is in specified gear or specified range is selected
- TPS within specified range
- EVAP is not purging
- VSS has been within specified range for specified amount of time without interruption

The following conditions will cause the Oxygen Sensor Monitors to be postponed:

Pending

- MIL is on and DTC is stored for:
  - Misfire
  - Upstream HO_2S
  - Upstream HO_2S heater
  - VSS

Conflict

- Certain segments of Fuel System Monitor are running.
- Engine has not been running long enough.
- Misfire detected, but no code stored yet.
- Upstream HO_2S failure detected, but no code stored yet.
- Upstream HO_2S heater failure detected, but no code stored yet.

Suspend

There are no suspend criteria programmed for the Oxygen Sensor Monitor because the test results are needed to enable other monitors.

DTCs and Action of the MIL

- DTC may be stored for one failure, no MIL.
- On the second trip with monitor failure, MIL is turned on and Freeze Frame data is captured.

- After three good trips, MIL will be turned off, but only if driving conditions were very similar to those when the DTC was set. DTC remains in memory.

Heated Catalyst and Catalyst Efficiency Monitor

Keeping a vehicle in compliance with emissions standards is the job of the Catalytic Converter. It is called a converter because it effectively converts harmful exhaust emissions to relatively harmless gases and water vapor. It does this by passing the exhaust stream across a heated catalyst bed inside the converter shell, which provides the chemical transformation. The catalyst must be kept clean and running in the correct temperature range to function. Keeping it working is one of the most important jobs the PCM has to do. By watching the oxygen sensors and tweaking the fuel delivery, the PCM maintains the catalyst at peak condition *if nothing goes wrong!* The Catalyst Monitor is designed to see how well the converter is doing its job. This monitor usually runs once per trip.

The catalytic converter, anywhere from one to four or more of them, is placed between two oxygen sensors. Certain models also may be equipped with mid-bed oxygen sensors, placed midway in the catalytic converter. The one between the engine and converter is called the upstream sensor and the one after the converter is called the downstream sensor. To monitor the efficiency of the catalyst, the Catalyst Monitor compares the readings of the upstream and downstream oxygen sensors. If a mid-bed oxygen sensor is used, its reading also is included in determining the efficiency of combustion and proper catalyst action. If the fuel delivery is correct and the catalyst is working efficiently, the reading at the upstream sensor will fluctuate back and forth from rich to lean. The downstream sensor will show much less fluctuation and the emission levels will be close to zero.

The enabling criteria for the Catalyst Monitor are:

- ECT at specified temperature (usually around 160°F, near complete warm-up)
- TPS indicates off-idle (part throttle)
- Engine operating in Closed Loop
- RPM and MAP or MAF within their specified range

The following conditions will cause the Catalyst Monitor to be postponed:

Pending

- Misfire DTC is stored.
- Any Oxygen Sensor DTC is stored.
- Downstream Oxygen Sensor DTC is stored for heater.

- Fuel Monitor DTC is stored for rich or lean.
- Engine running in Open Loop.

Conflict

- Insufficient time since engine start-up
- Any of the following monitors in process:
  - EGR
  - Fuel System, certain modes
  - EVAP Purge test

Suspend

- The Oxygen Sensor Monitor has not run with "passed" results.

DTCs and Action of the MIL

- DTC may be stored for one failure, no MIL.
- On the second or third trip with monitor failure, MIL is turned on.
- After three good trips, MIL will be turned off, DTC remains in memory.

EGR Monitor

Adding just the right amount of already-burned exhaust gases back into the intake system is the job of the EGR system, and it is a very exacting task, with little room for error. The key words here are "just the right amount." Too much EGR and the engine will lack power and "bog" under acceleration. Too little and the combustion chambers will run too hot, inviting detonation, aka spark knock. If the EGR flow is not correct, oxides of nitrogen (NO_x) emissions will not be properly controlled. As you might suspect, testing the EGR operation also requires some very precise procedures. Although there is no sensor available to directly measure the amount of NO_x emissions, auto manufacturers have invented several different types of monitors that allow the PCM to monitor whether the EGR system is working properly.

- Some models are equipped with a motor-driven EGR valve, which has a position sensor to measure how far the valve is open. The position sensor alone cannot determine actual EGR flow, only that the valve moved when commanded. The PCM will observe another input, such as MAP or oxygen sensor signals, to determine whether correct flow was achieved.
- Some other models use a vacuum-operated EGR, regulated by the interaction of a vacuum apply solenoid and a vacuum-venting solenoid. Again, an EGR

Figure 3–2   Photo of damaged DPFE sensor. Notice the corrosion present in the hose connections of this differential pressure feedback EGR sensor caused by heat, moisture, and acid.

position sensor is used to report how far the valve is open, and other inputs will be checked to determine that correct EGR flow is achieved.

- A temperature sensor may be installed in the EGR tube from the exhaust to the intake manifold. By measuring the temperature, the PCM can calculate the EGR flow.

- A Differential Pressure Feedback EGR sensor (DPFE) may be used to measure EGR flow by comparing pressure at two points along the EGR tube from the exhaust to the intake manifold.

It should be noted that, since the sensors mounted in the EGR tube are subjected to very high temperatures and a variety of exhaust gases and pressure pulses, they are somewhat prone to failure. Figure 3–2 shows a damaged DPFE sensor.

The EGR monitor is designed to use somewhat different testing strategies, depending on which of the measuring systems described above is used. However, most of the monitors test the EGR by operating the valve manually, outside of the normal, programmed position, and watching for the correct amount of change in the amount of EGR flow in response to the "open" or "close" command. Like the oxygen sensors, the EGR also is checked continuously for shorted and open circuits.

The typical enabling criteria for the EGR monitor are:

- Engine is warm (usually 180°F or more).
- Engine has been running for a specified amount of time.
- RPM, MAP or MAF, and TPS are within a specified range.
- VSS is within a specified range. (MAP/MAF, RPM, TPS, and VSS are usually in part-throttle cruising conditions for the EGR monitor to run. In these conditions, it will have a minimal effect on engine driveability).
- Short-term Fuel Trim (STFT) is within a specified range.

The following conditions will cause the EGR monitor to be postponed:

Pending

- EVAP monitor is running.
- Catalyst Monitor is running.
- The engine has not been running long enough.
- A misfire condition has been detected.
- Fuel System lean or rich failure has been detected.
- One or more Oxygen Sensor Monitor or Heater failures have been detected.

Conflict

- A misfire DTC is stored.
- Upstream Oxygen Sensor or Heater DTC is stored.
- Fuel System rich or lean DTC is stored.
- VSS DTC is stored.
- CKP or CMP DTC is stored.
- Engine is running in Open Loop.

Suspend

- The EGR monitor will be suspended until the Oxygen Sensor Monitor has run with passed results.

DTCs and Action of the MIL

- DTC will be set for a failure on one trip, no MIL.
- On second consecutive trip with same failure, MIL is turned on.
- After three consecutive good trips, MIL will turn off, but only if driving conditions were very similar to those present when the fault was detected.

EVAP Monitor

Evaporative emissions were not addressed by the OBD I generation of self-diagnostics. Leaking fuel storage systems and faulty devices such as liquid/vapor separators were not detected unless liquid fuel was spilling onto the ground. With OBD II, the evaporative emission controls are tested to make sure that the fuel system is preventing gasoline vapors from evaporating into the air. The switching valves and sensors also are monitored to be sure that the system can purge the accumulated vapors into the engine intake system.

Like the EGR system, the EVAP system is not the same on all makes and models. However, all must be leak free and have the capability to purge the accumulated vapors into the engine intake system at the right time. The EVAP monitor is run using several different test strategies, depending on the design of the specific system. The monitor may pressurize the plumbing and fuel tank, or it may apply a vacuum and watch for leaks to make sure they are airtight. It may operate the valves and watch for changes in the fuel system, usually by measuring the oxygen sensor readings. The monitor is programmed to "see" the difference between a small leak and a large leak in the EVAP system. It can also test the operation of every valve and sensor.

Generally, a series of combination vacuum switching valves and pressure sensors are located in the various lines, and sometimes in the fuel tank as well. They are used to check each section of the lines individually for leaks. Then, the valves are tested to make sure that the accumulated fuel vapors will flow on demand, as intended. This is accomplished by switching the valves open and closed individually while the monitor watches for an appropriate change in the upstream oxygen sensor readings and the resulting change in short-term fuel trim as it adjusts to the changes in oxygen content.

The enabling criteria for the EVAP monitor are:

- Engine running in Closed Loop.
- Oxygen Sensor Monitor has been run with passed results.
- ECT is 170°F or above.
- Transmission is in specified gear or selected range.
- TPS and RPM are within a specified range.

The following conditions will cause the EVAP monitor to be postponed:

Pending

- Engine RPM, TPS, or ECT move out of specified range before a given monitor step has been completed.
- Transmission is shifted to a different range or gear.

Conflict

- Oxygen Sensor Monitor is running.
- Fuel System Monitor is running.

Suspend

- Oxygen Sensor Monitor has not passed or has an active failure.
- Fuel System Monitor has not passed or STFT has an active failure.

DTCs and Action of MIL

- DTC will be set for a failure on one trip, no MIL.
- On second consecutive trip with same failure, MIL is turned on.
- After three consecutive good trips, MIL will turn off.

Secondary Air Injection or Air Injection Reactor (AIR) Monitor

The practice of injecting air into the oxidation bed of the catalytic converter and into the engine exhaust manifold is often thought of as old-fashioned, because it was used much more commonly in the earlier days of emission control systems. However, some vehicles equipped with OBD II still use this system to add oxygen to help the catalyst perform its job and to help burn excess raw fuel during warm-up before it reaches the catalyst. The differences are that the conventional belt-driven pump has been replaced by a self-contained, electric motor-driven pump on some models and that OBD II regulations require that a monitor be used to assure that the system is functioning, and its control valves are working properly. The AIR Monitor runs two basic types of tests to check the system's operation: the passive test and the active test.

The passive test measures oxygen sensor readings during warm-up, when the PCM has the secondary air directed upstream, into the exhaust manifold. As soon as the oxygen sensor is warm enough to function, the signal voltage should be low to indicate "lean," because of the air being pumped into the exhaust. Once the engine is warm and the engine management system switches to closed-loop operation, the PCM turns off the upstream air injection. Now, the upstream oxygen sensor should begin to toggle from rich to lean. If this change happens, the monitor passes.

If the passive test fails, or if the PCM doubts the results, it may perform the active test. For this test, the PCM switches on the upstream air injection temporarily after the engine management system is already in closed loop. Once again, the oxygen sensor signal voltage should change, this time to a low, "lean" reading.

Enabling criteria for the AIR Monitor are:

- Engine started and running in Open Loop
- Successful switching of engine management system to Closed Loop
- Valid upstream oxygen sensor(s) signal

The following conditions will cause the AIR Monitor to be postponed:

Pending

- Engine management stays in Open Loop.
- CCM has a pending DTC.

Conflict

- The Catalyst Efficiency Monitor is being run.
- The Oxygen Sensor Monitor is being run.

Suspend

- The engine is shutoff before Closed Loop is reached.
- Oxygen sensor DTC is present.
- Oxygen sensor heater DTC is present.

DTCs and Action of MIL

- DTC may be stored for one failure, no MIL.
- On the second trip with monitor failure, MIL is turned on.
- After three good trips, MIL will be turned off, DTC remains in memory.

All of the system monitors above are designed to detect problems in the engine systems and controls that may cause increased exhaust emissions as required for OBD II compliance. The only case in which any of the major monitors will not be present is where the engine is able to meet the established emission limits without installation of a particular system or component. For example, some vehicles do not need an EGR valve and/or secondary air injection.

Now that we have explored the details of OBD II monitors and their operating characteristics, let's move on to the business of diagnosis and troubleshooting.

Review and Reinforcement Questions

1. The specific combination of driving conditions that must occur in order to trigger a particular monitor is called a _____.

    a. Trip

    b. Drive cycle

    c. Warm-up cycle

    d. Key on, engine running (KOER)

2. If a monitor requires two trips to turn on the MIL, the DTC will never be stored on the first trip if a fault is detected. True or False?

    a. True

    b. False

3. HO_2S1, B1 has stopped functioning and is showing a continuous low voltage. Which of the following monitors will not likely be run?

   a. Catalyst

   b. EGR system

   c. EVAP system

   d. All of the above

4. All of the following monitors run continuously *except*:

   a. Secondary Air Injection

   b. Comprehensive Component

   c. Fuel System

   d. Misfire

5. Which of the following sensors are used by the Misfire Monitor?

   a. CKP

   b. Fuel Level

   c. MAP

   d. All of the above

6. The combination of driving conditions needed to allow all monitors to be run is called a_____.

   a. Trip

   b. Drive cycle

   c. Warm-up cycle

   d. Key on, engine running (KOER)

4

Diagnostic Trouble Codes and Scan Tools

Now that we understand monitors, we are ready to work with DTCs. Let's have a more detailed look at their structures and strategies and our troubleshooting strategies as well.

MIL Strategies

The MIL has a very important job to accomplish. While the auto repair technician has several interfaces available to provide diagnostic information, the MIL is the only OBD II interface with the vehicle driver. The MIL is used to indicate problems in the components and systems monitored by OBD II, and can appear in a variety of display formats, including "Check Engine," "Service Engine Soon," a symbol of an engine, or others. On some models, it is easy for the driver to confuse the MIL with other indicator lamps, such as the maintenance reminder lights used on many vehicle models. Sometimes, even the wording on the reminder lights is similar to the MIL, such as "Maintenance Required Soon," "Service EGR,"

"Service Required," or others. It is important to verify which indicator is actually turned on when a vehicle owner reports an illuminated MIL.

Auto manufacturers have a bit of flexibility in OBD II compliance with respect to the function of the MIL, but not much. According to federal law, the MIL *must* light for any fault that might possibly increase exhaust emissions by 50 percent or more above the levels measured in the new car Federal Test Procedure (FTP). The FTP is the test routine mandated by the EPA to certify emissions compliance for every new model car or light truck introduced.

In addition, manufacturers may elect to have the MIL light for faults that do not affect vehicle emissions at all. It is important to remember that an illuminated MIL does not necessarily mean an emissions-related fault. It simply means that the PCM has "seen" something that it is programmed to consider abnormal and is reporting the problem to the driver.

MIL Indications

To review some facts from Chapter 2, the MIL indicates problems in three different ways:

- If the MIL flashes quickly just once, it indicates a momentary fault. There is no need to seek repairs, and a DTC is not likely to be stored.

- A MIL that stays lit continuously while driving means that an ongoing fault has been detected. The DTCs should be diagnosed and repaired soon. In addition, the driveability and fuel economy may be affected.

- A flashing MIL indicates an urgent problem that can cause serious damage to the catalytic converter. The driver should reduce the vehicle speed and load immediately to attempt to stop the MIL from flashing. If it does not return to a "steady on" light, the engine should be shutoff, and the vehicle should be towed to a repair facility.

Two Types of Snapshots

When a DTC is set, there are two types of *snapshots* that may be captured and stored just as the fault occurs. This data can be very useful in troubleshooting those pesky intermittent failures, because the technician can see exactly which parameters were abnormal.

Freeze Frame data is a snapshot of generic emission information showing the condition of various readings at the time the DTC was set (not necessarily the first time the fault occurred).

A **Failure Record** is a snapshot of key data conditions at the time the fault last occurred. The Failure Record is updated each time the fault occurs. Not all manufacturers make use of the Failure Record feature.

Diagnostic Trouble Code Triggers

There are several ways in which a DTC may be triggered. In the DTC definition chart, some distinctions are made, but sometimes the reason for a code having been set will only be revealed as the technician performs the diagnostic routine. Three basic types of faults will trigger a DTC:

- A mechanical failure
- An electrical/electronic failure
- A rationality failure

An example of a mechanical fault would be no change in exhaust system pressure when the PCM commands the EGR valve to open, although the electrical circuits to the EGR actuator and the exhaust pressure sensor appear to be normal. It is still possible that the actuator or pressure sensor is at fault, but if so, the fault is likely to be mechanical in nature, such as a stuck actuator solenoid. It is also possible that a leak in the exhaust system or a disconnected hose to the pressure sensor is the cause. The OBD II system usually is capable of identifying a disconnected hose, because the exhaust pressure reading would be at or very near zero.

An electrical fault would occur in the same EGR system if the actuator or sensor circuit were open or shorted, or if the signal reading is outside of its normal parameter. The OBD II system is capable of telling the difference between these three electrical conditions.

A rationality fault is detected by the OBD II system's ability to compare input signals from several sensors to see if the information is logical. The readings must be within specifications, *and* they must make sense when compared to each other. If any input is not logical, the PCM will file a complaint in the form of a DTC. For example, OBD II will watch engine RPM as commands are sent to Idle Air Control (IAC). Changes in IAC must result in an appropriate change in RPM.

Diagnostic Trouble Code Classifications

There are four basic types of DTCs contained in the self-diagnostics of their OBD II–compliant vehicles, but only two of the categories are emissions related. This classification can help us to better understand the levels of urgency associated with different kinds of faults. Ford Motor Company, General Motors, and most import vehicles use the following classifications:

- **Type A codes** are emissions related, and are the most serious category. They are likely to damage the catalyst, or involve essential inputs to the PCM, usually ECT, MAF, MAP, TPS, and VSS. The MIL will light, and the DTC will be stored as soon as the fault is detected in most cases. A few Type A codes will not turn on the MIL until the second consecutive trip with the same fault. In addition, the Freeze Frame data will be captured and a Fail Record may be stored and updated each time the fault occurs.

- **Type B codes** are emissions related, but are less serious in nature. The DTC is recorded but not reported on the first trip with a failure. If the test passes on the next consecutive trip, the DTC will be erased. However, the MIL will be turned on, and the DTC will be stored if the fault occurs again on the second consecutive trip. Also on the second trip with the failure, Freeze Frame data will be captured and a Fail Record may be stored and updated each time the fault occurs.

- **Type C codes** are not emissions related. The MIL will not be turned on, but a DTC will be stored after the first trip with a failure. A warning lamp other than the MIL or a driver's display message may appear. No Freeze Frame data is captured. However, a Fail Record may be stored and updated each time the fault occurs.

- **Type D codes** are not emissions related. The MIL will not be turned on, and no DTC will be stored. A warning lamp or driver's display message may appear. No Freeze Frame data is captured. A Fail Record may be stored and updated each time the fault occurs.

DaimlerChrysler has a different classification system, known as code priorities. Chrysler DTCs are classed as follows:

- **Code Priority Type 1** This type of code is a one-trip failure of a two-trip code, which is *not* a fuel system or misfire failure. A Type 1 code also:
  - Stores Freeze Frame data (can be overwritten by a Priority Type 3 code)
  - Becomes a Priority Type 3 code on the second trip with same failure
  - Does not turn on the MIL

- **Code Priority Type 2** This class is a one-trip failure of a two-trip code, which *is* a fuel system of misfire failure. A Type 2 code also:
  - Stores Freeze Frame data (can be overwritten by a Priority Type 4 code)
  - Becomes a Priority Type 4 code on second trip with same failure
  - Does not turn on the MIL

- **Code Priority Type 3** A Type 3 code is the second failure of a Priority Type 1 code *or* the first failure of a CCM code. A Type 3 code cannot be overwritten by a Priority Type 4 code. A Type 3 code also:
  - Stores Freeze Frame data
  - Turns on the MIL

- **Code Priority Type 4**   A Type 4 code is the second failure of a Type 2 code *or* the first failure of a one-trip catalyst misfire code. A Type 4 code also:
  - ○ Stores Freeze Frame data
  - ○ Uses similar conditions of the first trip failure to decide whether or not to set a second trip failure
  - ○ Turns on the MIL
  - ○ Flashes the MIL upon catalyst-damaging misfire while it is occurring

It is important to check for stored DTCs as one of the first steps in your diagnostic routine, even though the MIL may not be on. The driver may report that the MIL was on, then turned off again, perhaps several days or even weeks ago. Unless the vehicle has been driven for a large number of drive cycles or the battery disconnected, odds are that the DTCs are still present in the PCM memory. If so, they can provide clues to help in chasing an intermittent problem, especially if Freeze Frame data also are available. It is a good policy to wait until you are finished with the repair before erasing the DTCs and data. You may want to refer to them as you continue your diagnostic routine. For example, the MAP, VSS, and TPS readings at the time of the fault can give you some idea of how the vehicle was being driven. You may want to duplicate the same conditions when road testing to verify the success of your repair. This can be done with pretty fair accuracy by noting the freeze frame readings, then trying to duplicate them using the scan tool's data stream mode as you drive the vehicle.

Structure of DTCs

OBD II generic DTCs are designed with a five-place alphanumeric structure that is quite easy to understand. The OBD II monitors are so sophisticated that the codes are highly detailed. For example, the PCM knows the difference between too little EGR flow and a disconnected EGR hose. Because of this ability to distinguish small differences in faults, there are over 4000 DTC numbers assigned *just for the powertrain!* Figure 4–1 explains how the DTCs are organized.

The first character identifies the broad system in which the trouble occurred: the Powertrain, Body, Chassis, or Network Communication.

The second character, X2, identifies whether the DTC being read is a generic OBD II code (ISO/SAE) or a manufacturer's code (OEM.) If the code is an OBD II generic DTC, "0," "2," or "3" is displayed. If it is an OEM code, "1" is displayed.

The third character, X3, identifies the system or subsystem that the DTC occurred in. Each range of code numbers is organized according to the general function they are related to.

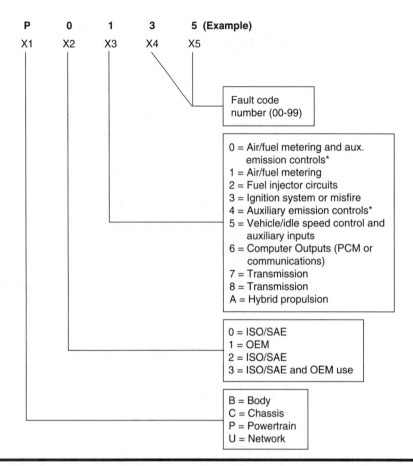

*Auxiliary emission controls generally include AIR, Catalyst, EGR, and EVAP. The first character, X1 above, identifies the vehicle area that the DTC is related to.

Figure 4–1 Structure of OBD II Diagnostic Trouble Codes.

The fourth and fifth characters, X4 and X5, are the numbers assigned to the specific fault within the system indicated by the third character.

There are 100 possible code numbers within each third character system category. Appendix C contains a more detailed list of specific powertrain DTC definitions. We will refer to these later, when we explain how to obtain the maximum benefit from DTCs.

Scan Tool Operation

The tool of choice for most OBD II diagnosis and repair is the scan tool. In nearly all cases, there is a diagnostic routine that will guide the technician toward pinpointing the cause of a symptom or DTC. However, the scan tool does have a few limitations. As we will see, other diagnostic equipment and techniques are sometimes needed. Sometimes, diagnostic shortcuts also can be found beyond the scan tool.

Data Display and Test Modes

SAE Document J-1979, Diagnostic Test Modes, outlines the operational modes of the OBD II scan tool in great detail. There are eight modes, each with a unique individual function. The diagnostic test modes are one of the features you should become familiar with in order to get the maximum benefit from OBD II diagnostic features. The diagnostic modes of operation are, as follows:

- Identification of what powertrain information is available to the scan tool, including sensor information, monitor status, and calculated values
- Oxygen Sensor Monitor and HO_2S test results
- Request for Freeze Frame data
- Total number of powertrain/emission DTCs stored
- Ability to clear the DTCs and Freeze Frame data
- Request for test results of noncontinuous monitors
- Request for test results of continuous monitors
- Request for control of on-board system, component, or test

Readiness Status

Previously, we have discussed standardization as one of the goals of the OBD II system. Yet, we have seen several instances where standardization really is not required and variations exist among manufacturers. Even within the same vehicle make, model, and year, different engines, designs, and options may cause "normal" parameters to vary considerably. Readiness status is yet another case where variations are allowed while still maintaining OBD II certification.

The Readiness Status monitor display on the scan tool provides a list of the monitors supported by the vehicle being tested. All OBD II certified vehicles have monitors for Misfire, Fuel, Oxygen Sensor, Catalyst, and Comprehensive Components. However, some vehicles do not have all of the monitors supported by OBD II. The usual reason for this is that a given vehicle may not be equipped with a particular component or system, such as Secondary Air Injection, EGR, or MAF. On most professional-grade scan tools, the readiness display will tell you which systems are installed and whether their respective monitors have been run. If the vehicle is equipped with separate computer/processors for engine, transmission, antilock brakes, body/accessories, and/or audio systems, the scan tool should indicate which monitor is being viewed. This is important, because certain information and sensor input may be shared by more than one system. The Readiness Status will not tell which monitors have passed or failed, only whether they have been run completely. In addition, the Readiness Status does not indicate how recently the monitors were completed, another feature that varies from one vehicle to another. It is important to remember that the DTCs and MIL indicate system faults, *not* the Readiness Status monitor!

Data and Parameter Displays

Earlier in this book we stated that the information supplied to the PCM has to be translated into a digital serial data stream, or "language" that it can understand. The same is true of the scan tool, in that it must translate the output serial data supplied by the OBD II system into a display the repair technician can understand. In effect, the data displayed on the screen of the scan tool is a translation of a translation. Although the information is generally very reliable, cases may arise where two scan tools will have a difference of opinion about what is being communicated. In some cases, different tools may even show different DTCs! That is why it is important to always follow a logical diagnostic and troubleshooting routine before attempting repairs.

Also important to remember is the fact that OBD II data communicated to the scan tool is not updated instantly. The minimum acceptable rate of transmission for generic OBD II data is 10 parameters per second, which is really quite slow. If your scan tool is set to display multiple parameters at once, which it almost certainly will be, the transmission rate may be even slower. The more parameters that are displayed on the screen, the longer the time between updates will be.

Sometimes, the technician may want to use a different diagnostic tool, which can read a signal directly rather than relying on the translation process. The tool of choice for this procedure is the Digital Storage Oscilloscope, commonly known as the *lab scope*. Once again, it is essential to have complete diagnostic software so that the information seen on the scope screen can be compared to known good pattern or voltage readings. Many high-end diagnostic tools contain both scan tool and DSO functions, as well as complete software for each.

DTC Diagnostic Strategies

Now let's look into the DTC testing strategies and how the technician can use them to guide troubleshooting and diagnostic procedures. First, we will discuss trips and verification of repairs. Then, we will explore several troubleshooting scenarios to illustrate the steps involved in getting the most from the DTCs. In studying the following content, be sure to remember this:

The best single diagnostic tool you can get is your thorough knowledge of how the systems and components monitored by OBD II function. If you understand what is supposed to happen, it is much easier to understand what the OBD II monitors and data are telling you.

Trips and Verification of Repairs

One of the best ways to verify successful repair of a problem that caused a DTC to be set is to have the PCM turn off the MIL without erasing the DTC on the scan tool. In addition, understanding the process of setting the DTC will aid in the troubleshooting process. To accomplish this, we must first understand the significance of the trip and the drive cycle.

In a general sense, the purpose of a *trip* is to tell the PCM what to do with the data it receives. For example, if the MIL is not on and the vehicle is not driven to meet the enabling criteria for a particular monitor, the monitor simply will not be run. If it doesn't run, no failure will be recorded, the MIL will not be turned on, and no DTC will be set, no matter how far out of "normal" the readings sent to the PCM may be. This is why the Readiness Status monitor display is important. If the technician does not look to see what monitors have been run to completion, important clues to guide the troubleshooting process may be missed. If a particular monitor can be ruled out because it has not been run, more valuable information may be obtained by using the data stream or PID displays.

Refer to the descriptions of the major monitors in Chapter 3. It is very important to remember that not all monitors use the same definition of a "trip," because of the differences in enabling criteria and whether one, two, or more trips are required to report a fault. Once again, different manufacturers may have different definitions of a trip for a given monitor. In addition, there are substantial variations in the drive cycles required from one manufacturer to another, one vehicle model to another, and even one monitor to another on the same vehicle. It is important to remember that one drive cycle does *not* fit all! You must refer to your service information. Look up the exact drive cycle requirements for the exact vehicle and the exact monitor you wish to trigger. This is the only way to be sure that you will manage to complete a trip that counts.

Despite the vagaries of trips and drive cycles, the MIL will be an accurate indicator of successful repair on a given vehicle. By logging trips "by the book," using the exact enabling criteria for a particular monitor, the technician can actually use the monitor to turn off the MIL. If three trips are recorded as completed with the enabling criteria being met, the MIL will turn off *in most cases*. In virtually all cases, however, the operating conditions contained in the OBD II Drive Cycle for a given vehicle and monitor should provide enabling criteria for the monitors being tested and store the individual results in memory. Three consecutive drive cycle "passing" grades should turn off the MIL. In addition, the DTC will be erased after a certain number of trips with no failure are logged. Typically, 40 trips are the number needed to "throw out" a stored DTC. As you can see, the DTC stays around for quite some time after the MIL is no longer lit.

As you might guess, the specific Drive Cycle conditions vary from one year, make, and model to another. It is critical to have complete and accurate service manual information available to make sure that you are meeting the exact conditions specified by the manufacturer when driving the vehicle. Remember, the enabling criteria must be met and the monitor must be run to completion or it does not count. Also, remember that more than one trip with a failure may be required to turn on the MIL. Do not assume that the problem is repaired just because the MIL does not come on with a short test drive. Verify the repairs by repeating the drive cycle, or check for pending DTCs. In fact, it is a good idea to do both! To be sure that all problems have truly been solved, road test the vehicle using the manufacturer's global drive cycle, if one is available for the vehicle you are servicing. Repeat the global drive cycle twice, then check for pending DTCs with the scan tool.

Examples of Global Drive Cycles

The following are examples of typical manufacturer-specific OBD II global drive cycles, which will provide complete setting of monitors and readiness status checks. In each case, the vehicle makes and models the drive cycle applies to are shown in order to illustrate differences and similarities from one manufacturer to another. The steps involving preconditioning and reset procedures using the scan tool have been omitted.

Example One: 2003 BMW 325xi

1. Engine cold start, idling, approximately 3 minutes. Evaluates:
    a. Secondary Air System
    b. Evaporative Leak Detection
2. Constant driving at 20 to 30 mph, approximately 4 minutes. Evaluates:
    a. O_2 Sensors Achieved Closed Loop Operation
    b. O_2 Sensors Response Time and Switching Time
3. Constant driving at 40 to 60 mph, approximately 15 minutes, including vehicle coasting phases. Evaluates:
    a. Catalytic Converter Efficiency
    b. O_2 Sensors Response Time and Switching Time
4. Engine idling, approximately 5 minutes. Evaluates:
    a. Fuel tank leak diagnosis (certain vehicles only)

Note   The diagnostic sequence will be interrupted if:

- Engine speed exceeds 3000 rpm
- Vehicle speed exceeds 60 mph
- Large fluctuations in the accelerator pedal position occur

Example Two: 2000-2004 Ford, Lincoln, and Mercury Vehicles

1. Idle for approximately 15 seconds. Drive at 40 mph until ECT is at least 170°F.

> **Note** If IAT is not within 40 to 100°F, EVAP monitor must be bypassed as shown in step 9 below. Evaluates:

   a. Engine warm-up

   b. IAT input to the PCM

2. Cruise at 40 mph for at least 5 minutes. Executes the O_2 sensor Monitor.

3. Cruise at 45 to 65 mph for 10 minutes, avoiding sharp turns and hills. Avoid wide-open throttle. Fuel level in tank must be between 15 and 85 percent. Executes the EVAP monitor if IAT is as specified above.

4. Drive in stop-and-go traffic conditions, including five different constant cruise speeds, ranging from 20 to 70 mph, over a 10-minute period. Executes the Catalyst Monitor.

5. From a stop, accelerate to 45 mph at 1/2 to 3/4 throttle. Repeat three times. Executes the EGR monitor.

6. Bring vehicle to a stop. Idle with the automatic transmission in Drive or manual transmission in Neutral for 2 minutes. Executes the ISC portion of the CCM.

7. For manual transmission, accelerate from 0 to 50 mph, continue to next step. For automatic transmission, place selector in "Overdrive." Moderately accelerate from 0 to 50 mph and cruise for at least 15 seconds. Stop vehicle and repeat without overdrive, accelerating to 40 mph and cruising for at least 30 seconds. While cruising at 40 mph, activate overdrive and accelerate to 50 mph, cruise for at least 15 seconds. Stop the vehicle for at least 20 seconds. Repeat step 7 five times. Executes transmission portion of the CCM.

8. From a stop, accelerate to 65 mph. Decelerate at closed throttle to 40 mph without applying brakes. Repeat this step three times. Executes learning for the Misfire Monitor.

9. If EVAP bypass is necessary (see step 1 above), park vehicle for a minimum of 8 hours. Repeat steps 1 through 8. This allows the "bypass" counter to increment to "2."

Example Three: 2001-2003 GM Vehicles, Except Metro, Prizm, Saturn, and Tracker

1. Cold start, coolant temperature less than 122°F. Turn off all accessories, idle for 2 minutes with transmission out of gear. Monitors O_2 Sensor Heaters, Misfire, Secondary Air, Fuel Trim, EVAP Purge.

2. Accelerate at part throttle to 55 mph, cruise at this speed until engine reaches operating temperature (may require 8 to 10 minutes, depending on start-up

temperature). Continue cruising at 55 mph for an additional 6 minutes. Monitors Misfire, Fuel Trim, EVAP Purge, O_2 Sensors.

3. Reduce cruise speed to 45 mph, maintain for 1 minute. Decelerate from 45 mph for 25 seconds with closed throttle, no brakes applied, clutch engaged with manual transmission. Allow vehicle speed to reach no less than 25 mph. Return vehicle speed to 45 mph at part throttle and maintain for 125 seconds. Repeat for four decelerations. Monitors EGR, Fuel Trim, EVAP Purge.

4. Accelerate at part throttle to 45 to 55 mph, maintain for 2 minutes. Decelerate to stop. Idle for 2 minutes with brakes applied, automatic transmission in Drive or manual transmission in Neutral with clutch pedal depressed. Monitors Catalyst, Misfire, EGR, Fuel Trim, O_2 Sensors, EVAP Purge.

Example Four: 1997 Nissan Altima XE

This manufacturer has identified three *patterns* within the global drive cycle for this vehicle. The patterns must be successfully completed in order, or the next pattern will not run the OBD II monitors properly.

1. Pattern A: Start engine and idle or drive vehicle until coolant temperature reaches at least 158°F. Turn off engine. Enables Misfire and Fuel System monitors.

2. Pattern B: With engine warm, start and idle briefly. Then, drive vehicle up to legal freeway speeds for several minutes. Return to idle and shutoff engine. Enables EGR and fuel system component (CCM) monitors.

3. Pattern C (for repair verification): Check scan tool Freeze Frame data to determine what speed and load conditions were present when failure occurred. Drive vehicle to recreate these conditions. For catalyst monitor, a steady cruise of 2 minutes or more should be achieved. Enables catalyst monitor and tests other components/systems as indicated by Freeze Frame data.

The manufacturer recommends repeating each of the specified driving steps three times. The specific drive cycle steps are:

1. Cold start-up, coolant temperature is 14 to 95°F, fuel temperature sensor is greater than 32°F. Idle engine until coolant temperature reaches 158°F.

2. Use the scan tool to indicate vehicle speed. Accelerate vehicle to 56 mph, then release accelerator and decelerate with closed throttle for 10 seconds. After 10 seconds has passed, return to 56 mph and hold throttle steady for approximately 6 minutes.

3. Stop vehicle and remain stopped for 10 seconds. Accelerate to 30 to 35 mph within 10 seconds and maintain speed for 20 seconds. During acceleration, hold throttle steady at part throttle (0.8 to 1.2 volts on TP sensor as indicated by scan tool). Repeat this pattern at least 10 times or until the EGR System Readiness Test (SRT) is set as indicated by the scan tool.

Example Five: 2000 Volvo, All Models

In this example, the manufacturer does not specify which monitors are being run in each step. However, the sequences are similar to those in other examples above. If a scan tool is used to monitor the System Readiness checks, changes may be observed as the trip cycles are completed.

Drive cycle steps:

1. Start and run the engine until the coolant is at least 149°F. Then, increase engine speed to approximately 2000 rpm for at least 5 minutes.

2. Stop the engine. Let it cool until ECT is between 39 and 104°F.

3. Switch the ignition On for at least 2 seconds.

4. Start the engine when the ECT is between 39 and 104°F. Move the gear selector to Drive (D) and accelerate gently to 40 to 50 mph. Drive gently for 5 minutes at 40 to 50 mph maximum.

5. Bring the vehicle to a stop. Allow the engine to idle for 30 seconds.

6. Accelerate gently to 40 to 50 mph. Drive gently for 5 minutes at 40 to 50 mph maximum.

7. Bring the vehicle to a stop. Allow the engine to idle for 30 seconds.

8. Accelerate at 50 percent throttle to 40 to 50 mph. Drive at a constant speed for 5 minutes at 40 to 50 mph maximum.

9. Bring the vehicle to a stop. Allow the engine to idle for 1 minute. Turn ignition off.

Examples of Specific Repair Verification Drive Cycles

Another means of verifying most emission and driveability repairs is to use the manufacturer's drive cycle that tests only the component or circuit that was repaired. Depending on the nature of the repair, the drive cycle may involve enabling one or more OBD II monitors, or it may be a simple road test to determine that the repaired item is now functioning correctly. The obvious advantage of this method is that it generally saves a substantial amount of time compared to the global drive cycle. The drawback is that not all monitors or system readiness tests will be completed, so the overall engine management picture will not be seen.

The following example shows a specific repair verification test:

Vehicle: 2001 Lexus LS 430

Repair Completed: DTC P0130/P0150: Heated Oxygen Sensor Circuit Malfunction.

Verification Test Procedure:

1. Connect scan tool to DLC No. 3. Start and warm engine to normal operating temperature.

2. Allow engine to idle for 100 seconds or more.

3. Drive vehicle for 20 seconds at 24 mph or more.

4. Stop engine and allow to idle for 20 seconds.

5. Repeat steps 3 and 4 two more times. At least idle test, allow engine to idle for 30 seconds.

6. If using OBD II generic scan tool, turn engine off. Repeat steps 1 to 5.

Getting the Most from DTCs

One of the advantages of OBD II compared to earlier self-diagnostic systems is its ability to provide detailed information about any fault it detects. Not only does the PCM detect a problem, it knows the difference between a sensor malfunction, a response issue, and a circuit that is shorted or open. Refer to the chart of DTC definitions in Appendix C, and look at the series of Oxygen Sensor DTCs, code P0130 through P0135. All of these codes refer to Oxygen Sensor 1 (upstream), Bank 1. You will note that specific definitions are listed for O_2 Sensor malfunction, O_2 Sensor voltage low, O_2 Sensor voltage high, O_2 Sensor slow response, O_2 Sensor circuit inactive, and O_2 Sensor heater malfunction. With OBD I, any of these six faults would likely have been recorded as a single DTC. OBD II streamlines the troubleshooting process by indicating exactly *how* the component or system failed. The technician does not have to waste time checking the signal from the O_2 Sensor if the fault was detected in the heater.

Now, look at the definitions for the ECT in Appendix B, P0115 through P0118. Once again, there are four different problems identified for this one sensor. This sensor is one of the components covered by the CCM. The "Range/Performance" DTC, P0116, indicates that the sensor is functioning, but may be reading incorrectly compared to the data of other sensors.

To use the DTCs to their maximum advantage, the technician should begin by looking at the *complete, exact description* provided in the DTC definition chart. As illustrated above, some DTCs will identify the exact circuit or component that failed, and may indicate what the PCM saw wrong with it.

Freeze Frame Data

Once the DTC definition is understood, it may be possible to verify what the PCM saw by examining the Freeze Frame data. The Freeze Frame is a single frame snapshot of serial data indicating exact engine operating conditions at the moment the DTC was set. The data stored typically includes:

*   Open or Closed Loop fuel system status*
*   Engine RPM

- ECT
- IAT
- Vehicle speed (VSS)
- Misfire data (may indicate specific cylinders misfiring)
- MAP (intake manifold vacuum)
- Calculated engine load (as a percentage)
- Fuel pressure (if adjusted by the PCM)
- STFT and LTFT (as a percentage)*
- Upstream and downstream HO$_2$S voltage*
- The DTC that triggered the freeze frame capture
- The total number of DTCs stored

Specific manufacturers may elect to make more data available in a freeze frame snapshot. Sometimes, the additional data can be viewed in generic OBD II diagnostic mode. However, it must be viewed in the OEM diagnostic mode in some cases.

The Freeze Frame data display is a very valuable diagnostic tool, because it allows the technician to compare all of the data captured to see what may be related to the exact failure defined in the DTC. By doing this, the exact troubleshooting steps to follow are more easily identified and much time can be saved in the diagnosis and repair processes. In addition, the exact causes of intermittent failures are more easily discovered. The combination of Freeze Frame data also is useful as a reference in achieving *similar conditions* during road test drive cycles to verify that repairs were successful. When enabling criteria have been met and the scan tool data stream shows similar readings to those captured in Freeze Frame, the same DTC should be set again if the problem is not repaired.

Erasing the DTCs

A very important point to remember is that DTCs, Readiness Status indicators and Freeze Frame data are the only details the technician can get from the PCM about what was detected by the OBD II system, and they all provide valuable diagnostic clues. When you erase the DTCs, you also erase the Freeze Frame as well as resetting at least some of the Readiness Status monitors. Once reset, at least one drive cycle, probably more, will be required to reestablish the monitors that have been run to completion.

It is good practice to always look at all of the data you can gather from the DTCs, Freeze Frame, Readiness Status monitors, and data stream. Evaluate the data

---

*These items are displayed separately for each bank of cylinders.

carefully and use it to pinpoint the conditions that caused the fault before you even consider erasing DTCs. Erase the DTCs only when you are sure that you found and repaired the root cause of the failures detected.

The old-fashioned idea of erasing the DTCs and waiting to see which, if any, reset on subsequent trips is generally *not* a good diagnostic strategy. The only time to apply this technique is when there is strong evidence that the DTCs stored are unrelated to the problem at hand and may have been caused during the process of someone else's diagnosis attempts. Even then, it is a good idea to print or write down the DTCs and also the Freeze Frame data. Attach the noted information to the shop's copy of the repair order so it can be referred to later if needed.

In the next chapter, we will explore several types of diagnostic and troubleshooting routines. In addition to using the information above, we will learn how to use the scan tool parameter and data stream displays and how to perform tests on individual circuits and components.

Review and Reinforcement Questions

1. The scan tool will not communicate with an OBD II vehicle. Technician A says to check the ground at DLC pin 4. Technician B says to check the resistance between the PCM case and chassis ground. Who is right?

    a. A only

    b. B only

    c. Both A and B

    d. Neither A nor B

2. HO_2S1, B1 refers to which location?

    a. Cylinder Bank 1, upstream of catalyst

    b. Cylinder Bank 2, upstream of catalyst

    c. Cylinder Bank 1, downstream of catalyst

    d. Cylinder Bank 2, downstream of catalyst

3. "Bank 1" refers to the line of cylinders:

    a. Closest to the firewall

    b. Closest to the front of the engine compartment

    c. Including the cylinder closest to the front of the engine

    d. Including the cylinder closest to the distributor

4. The PCM is able to adjust fuel trim to compensate for all of the following conditions *except:*

   a. A minor intake manifold air leak

   b. Slight clogging of the fuel injector nozzles

   c. Substantially lower than normal fuel pressure

   d. EGR flow slightly lower than normal

5. When the MIL illuminates briefly, then goes out, it indicates:

   a. An intermittent failure

   b. A DTC is definitely stored

   c. The failure is not emission-related

   d. All of the above

6. If you erase stored DTCs with a scan tool, Freeze Frame data will remain stored for 40 warm-up cycles. True or false?

   a. True

   b. False

5

Diagnostic Troubleshooting Routines

In this chapter, we will explore general troubleshooting routines and how to apply them to problems detected by the OBD II system. We will discuss specific scan tool usage, as well as testing techniques to find and verify problems indicated by the scan tool.

General Troubleshooting Steps

The following general steps will be helpful to put in perspective all of the step-by-step "decision tree" type of diagnostic routines that the technician often encounters in troubleshooting an OBD II DTC. These steps will apply to virtually any malfunction. The eight steps to success are:

1. Gather information
2. Look and listen
3. Five-minute quick checks
4. Read the DTCs
5. Check for service updates
6. Observe all data available
7. Perform specific DTC routine(s)
8. Verify repairs

Step One: Gather Information

No matter what type of diagnostic testing equipment will be subsequently used, the first step in troubleshooting is to gather all of the information you can before you lay hands on the vehicle. If possible, talk to the owner to determine a logical starting point for your diagnosis. Some things you will want to know:

- Are there any driveabiltiy symptoms, or just an illuminated MIL? If so, when do the symptoms occur and exactly what happens?
- How long have the symptoms and/or MIL been present?
- Does the MIL flash off and on at certain times while driving?
- Has the vehicle been serviced elsewhere for this same problem? If so, are copies of the repair orders available?
- Have the manufacturer's recommended service procedures been performed at the right intervals?

Considering all of the information and concepts we have covered up to this point, it is easy to see that using the owner as a source of information may help to determine whether the problem that turned on the MIL is likely to be a mechanical malfunction, an emission-related fault, or something that may be corrected by performing needed routine maintenance procedures. For example, a poorly maintained vehicle may be in obvious need of spark plug or plug wire replacement, which in turn, has triggered one or more misfire DTCs.

Step Two: Look and Listen

Once you have gathered as much information as possible from the owner, begin your troubleshooting by inspecting the vehicle thoroughly. Even if you do not think there is a relationship to the problem, check the basics. Visually inspect the condition and levels of all fluids. Remove the radiator cap and verify that the cooling system is full if the engine is cool enough to do so safely. If not, put this on your list of things to do later, after it has time to cool down. Look at the condition of the exhaust system and check it for leaks. Under the hood, look for vacuum and emission control hoses that are cracked, collapsed, broken, disconnected, or missing altogether.

See that all engine and accessory components are present and properly installed. Also, make a note of any components or systems that appear to be modified. Don't assume that this is the cause of the problem, at least not until you have read the DTCs and established that there is a likely relationship.

Start the engine and observe the idle quality. Look and listen for misfires, unusual noises, or other abnormal conditions. Walk around the vehicle and listen for excessive fuel pump noise or the rattling sound of a loose, likely burned out, exhaust catalyst bed. If there are driveability symptoms that could be related to incorrect fuel pressure, you should test the fuel pressure to the rail and make sure that it is within specifications. On some vehicles, it is possible to monitor the fuel pressure with the scan tool, using the data stream display, but many models do not have this feature.

It may seem to be a waste of time to perform basic inspections before hooking up the scan tool and reading data. However, this technique, along with the other basic tests to follow in the section "Step Three: Five-Minute Quick Tests," are *monitors* that you are running before consulting with the PCM. You may discover obvious problems that will need to be repaired before the DTCs and scan tool data can be considered accurate. If you follow the basic troubleshooting steps recommended in this chapter, you will be surprised how many times a problem that seems to be "high tech" in nature is often caused by a very basic mechanical problem or a component that has been tampered with.

Step Three: Five-Minute Quick Tests

If the vehicle passes all of your visual inspections without the need for correction of basic problems, it is time to perform some quick tests to determine the integrity of the main electrical components and circuits. Most sensor signals to the PCM are based on measurements compared to a reference voltage applied to them. This measurement requires proper, steady battery voltage and good ground connections, or the signals may be inaccurate or completely out of their normal range. A basic electrical problem should always be suspected when a number of DTCs relating to sensor signals or PCM output failures are stored. It is important to remember that the battery is the vehicle's source of electrical power, so its function interrelates with virtually all electrical and electronic components. Also, remember that the grounds are the pathway for electrical current to flow from the battery to all of the components it supplies. It is critical in the operation of any electrical or electronic component that both battery and grounds are functioning as designed.

Begin this step in the troubleshooting process by checking the battery condition and state of charge. If you have a good quality battery tester available, it is best to use it to test both the battery and charging system. If not, you can test the battery's state of charge and the charging system voltage with a voltmeter. An inexpensive **digital volt/ohmmeter (DVOM)**, also known as a **digital multimeter (DMM)**, is all that is needed. When shopping for a DMM, be sure to obtain one that has an *input impedance* of 10 megaohms or more. This specification will assure that the meter will not draw enough current in taking its readings to damage the sensitive electronic circuits you may be measuring with it. A high-impedance meter also is very accurate. As you will see a bit later on, accuracy is very important, since some "okay ranges" for sensors are as little as 0.3 volts or even less.

Start the electrical tests by assessing the battery's state of charge. The easiest and quickest way to do this is by performing the Open Circuit Voltage (OCV) test. As the name implies, we are going to test the battery with all circuits open, or at least as close as we can get to that condition! Turn off all accessories, lights, etc., remove the key from the ignition, and close the doors. See that the interior lights are turned off. Don't forget to unplug the under-hood light or remove the bulb from it if the vehicle is equipped with one. The idea is to come as close as we can to having no current flowing through the battery while we perform this test. Now, turn the headlights on for 15 seconds to remove any surface charge from the battery. Then, turn them off again. This provides a more accurate test reading.

Once everything is turned off, simply measure the voltage across the two battery terminals with your DMM. Many people are surprised by the small difference in OCV between a fully charged and a discharged battery. Figure 5–1 shows the state of charge indicated by the OCV test.

Since each cell in a fully charged battery produces about 1.2 volts, a reading in the 10-volt area indicates the likelihood of a dead cell. Replace the battery before proceeding with more diagnosis. If the OCV reading indicates less than a 75 percent state of charge, there may be a problem with the charging system or a defect in the battery. Either can cause skewed sensor inputs and cause false OBD II fault detection. It is important to remember that OCV *only indicates the battery's state of charge, not its overall condition!* We will check the battery's capacity a little later on.

To test the charging system, leave the DMM connected to the battery terminals and start the engine. Observe the charging voltage. The rule of thumb is that the charging voltage should be at least 0.5 volts above the OCV of the battery, but check the manufacturer's specifications to be sure the charging voltage is acceptable. Turn on all of the electrical loads, such as headlights, A/C blower, and rear window defogger. Now, check the charging voltage again. It may be normal for the voltage to fall close to the OCV reading at idle, but it should rise back to its normal range if the engine RPM is increased to around 1500.

To complete the charging system tests, check for the presence of too much *AC ripple* in the charging system current flow. Ripple is the amount of alternating

The following chart shows the state of charge indicated by measuring the voltage across the terminals of a 12-volt automotive battery with all electrical accessories turned off. It is important to remember that open circuit voltage indicates only state of charge, not battery condition. If a battery shows a low state of charge, it may be assumed that the weak battery is the most likely cause of a slow-crank or no-crank condition. If the battery indicates 75% or greater state of charge, the battery may be defective or the problem may be elsewhere.

Open Circuit Voltage	State of Charge
12.6 or above	100%
12.4	75%
12.2	50%
12.0	25%
11.8 or less	0%

NOTE: Each cell is 2.1 volts when charged. So, ~10 volts would indicate a "dead" cell.

Figure 5–1 Quick Check Chart 2: Open Circuit Battery Voltage.

current that the alternator allows to "slip through" in charging the battery. To test for AC ripple, leave the DMM connected to the battery terminals and switch from DC volts to AC volts. Choose a low scale on the meter, preferably a 0- to 2-volt range. Observe the AC voltage reading with the electrical loads turned on and again with them turned off. In either case, a reading of over 0.125 volts indicates that the alternator diodes are *leaking* alternating current into the vehicle's direct current system. Excessive ripple can cause numerous problems and even damage sensitive electronic components, including the PCM and its OBD II system functions.

After the battery OCV and charging system have passed your monitors, take 1 minute more to check the starting system. This test actually has two purposes. First, it provides a crude type of capacity, or "load" test for the battery, something we cannot do directly with just a DMM. Second, it tells whether the battery has sufficient power to start that particular engine without losing too much of its voltage. This is very important because very low voltage levels while cranking can cause PCM inputs and outputs to send false readings, and can even cause PCM memory loss in some cases.

To quick-check the battery capacity and starting system, leave the DMM connected to the battery terminals and set to read the DC battery voltage. Then, crank the engine and observe how far the battery voltage drops while it is cranking. If the engine starts too quickly to obtain a reading, you may have to unplug the ignition coil or module, or remove the fuel pump fuse to prevent it from starting. During the

Battery Temperature °F(°C)	Minimum Voltage Cranking/Load Test, volts
70 (21) or above	9.6
60 (16)	9.5
50 (10)	9.4
40 (4)	9.3
30 (−1)	9.1
20 (−7)	8.9
10 (−12)	8.7
0 (−18)	8.5

Figure 5–2 Minimum Battery Capacity Varies with Temperature.

cranking voltage test, the battery voltage reading should stay above 9.6 volts at 70°F or above. Refer to Figure 5–2 to see the effect that temperature has on the minimum acceptable battery capacity. If cranking voltage is too low, the battery is weak or the starter is drawing too much current. The only sure way to tell what is wrong is to do a complete battery test with the proper equipment.

Finally, test the quality of the vehicle's main ground connections with your DMM by performing a voltage drop test. Connect the DMM to the battery's ground terminal, directly to the post if possible, and to an unpainted area on the engine. Set the meter on a low DC voltage scale—0 to 2 volts is a good choice. Now, crank the engine and observe the voltage reading. Start the engine, turn on all of the loads, and observe the voltage reading again. If the voltage reading remains less than 0.5 volts, the ground quality at the chosen test point is good, as illustrated in Figure 5–3. Now, move the DMM lead from the engine to an unpainted area on the body. If possible, connect to a point where ground wires are attached to the firewall, wheel well, or strut tower. Repeat the test. Once again, there should be very little voltage drop as the electrical loads are applied. If the voltage drop is excessive, clean or repair the ground connections before proceeding with further diagnosis.

Never overlook the importance of testing the basic power sources and quality of electrical connections. A poor ground has left many a technician with a repair job that became a nightmare and seemed to turn into "mission impossible!"

Step Four: Read the DTCs

Once you have performed your monitors and eliminated the obvious and the basics, it is time to bring the PCM into the troubleshooting process. Plug in the scan tool and read the DTCs, check the Freeze Frame data, and evaluate the information obtained from them as we have previously discussed. Look at the data

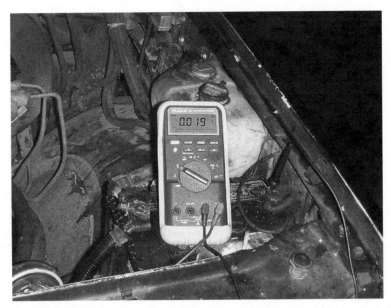

Figure 5–3 A good connection! Voltage drop of the main ground is less than 0.2 volts. No problem here.

parameters. On this "first pass" over the readings, look for anything that is "way off" or a parameter that appears suspicious. Then, see if the suspect information relates to a component or function that is tied to the DTC. In many cases, you will identify what seems to be a clear-cut cause and a definite diagnostic path at this point. If so, the next step is to test the suspect component or circuit to verify your hunch. Proceed to Step Five.

> **Note** In the process of reading the scan tool information, remember to consider its possible relationship to any abnormal items discovered in Steps one and two. It is important to think of basic root causes that may show up on the scan tool data as an apparent sensor or circuit failure. For example, a leaking or kinked MAP sensor hose will cause the same abnormal reading as a defective internal circuit in the sensor itself.

Step Five: Check For Service Updates

If Steps one through three did not provide you with a definite path to follow with your diagnosis, take time out from hands-on troubleshooting and look for updated service information from the manufacturer, which may shed some light on a perplexing problem. You may find that you have been chasing a "ghost," in the form of a glitch in the PCM programs. This approach is especially helpful if the vehicle you are diagnosing has an illuminated MIL, but no driveability concerns. Sometimes, the PCM's software simply needs to be updated with the latest programming from the manufacturer to correct glitches that may not have shown up until after the vehicles were released for sale. This type of malfunction is generally found on nearly new vehicles with low odometer readings, but automobiles have a habit of doing the unexpected! The possibility of an update is worth considering before you continue your troubleshooting.

If you have access to an automotive service and repair database, such as one of the online subscriptions, check for technical service bulletins or other service updates to see if the vehicle you are working on is eligible for any reprogramming, recalibration, component replacement, or other updates that could be related to the DTC you are working with. If no direct source of information is available, check with the local dealership service department to see if they show a "fix" or history of the problem you are experiencing. Be advised that, just because a manufacturer has identified a concern and published a correction, it does not mean the repair will be free of charge at the dealership. Technical service bulletins and updates generally are not recalls and are usually not free unless the vehicle is still covered under its new car warranty.

Another Source for Gathering Information

We would be remiss if we did not mention the availability of manufacturers' service information on the Internet, other than through a paid subscription. Of course, you will still need a modem or broadband Internet connection, as well as an Internet Service Provider (ISP) to access and download what is out there. One rather "central" site is a federal government collection of information called FedWorld. The Internet address is www.fedworld.gov. Each manufacturer has placed a lot of individual files including training materials, technical service bulletins and special tools in large, downloadable files, which are compressed, or "zipped" to reduce the amount of memory required to store them.

The key word here is "large." If you are working with a dial-up connection to the Internet, trying to download these monster files can be frustrating at best and cause you to be "booted off" by your ISP at worst. They will also eat up a lot of memory on your computer. The best advice we can provide is to go to the site and sniff around to see what you can find and use. The good news is that there is a main index of files that you can download and keep handy for future reference. Unfortunately, this site is not user-friendly. Once you unzip a file to view it, the manufacturer's index is no longer present. Therefore, you have to open the individual files to view them, which is both tedious and time-consuming. Online service information subscriptions are expensive, but they are much quicker and friendlier.

Also on the Internet, you may find technical service bulletins and useful update information, such as emergency service procedures and information about the latest hybrid power plant–equipped vehicles or other news by going straight to the individual manufacturers' Web sites. Another approach is to go to a forum or "bulletin board" site where other auto technicians may be available to help with a specific problem on a particular vehicle. One such site is the International Auto Technicians Network (IATN). At any given time, about 1800 of your fellow technicians will be reading your question or concern. If any of them have been in the same situation and found a solution, they are pretty quick to offer help. You will also find that you can contribute your knowledge to help them with their problem jobs. It's a win-win proposition!

Step Six: Observe All Data Available

The next step in the troubleshooting and repair process is to pinpoint the root cause of the problem. The beauty of OBD II DTCs is that most of them are very specific. In most cases, the DTC will tell what component or circuit has failed, as well as how it failed. That feature streamlines the testing portion of the diagnostic routines considerably, compared to earlier OBD I systems.

Begin your hands-on verification of the indicated failure by identifying the specific system, subsystem, or circuit indicated by the DTC. You may need to consult a wiring diagram or service manual to know what color wires to test or exactly where a particular component is located. Once you know that information, consult your service information source (scan tool diagnostic software, repair database, or service manual), to find out what to test and what the test results should be. You must look up each vehicle make, model, and year individually.

Next, check the parameter ID display on the scan tool. Print out or record the parameters for the system and components indicated in the DTCs. This step will give you a quick reference to used in the more detailed DTC diagnostic routines to follow. Do not trust general normal ranges of sensor operation, as the particular vehicle you are working on may be very different! Figure 5–4 shows a typical PID display on a scan tool.

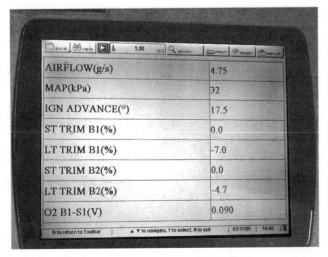

Figure 5–4 Scan tool PID display.

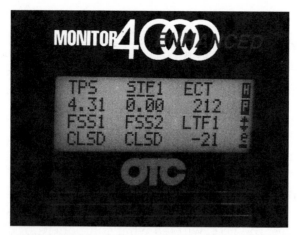

Figure 5–5   Scan tool data stream display.

Now, switch the tool to the data stream mode, as illustrated in Figure 5–5. Check the data relating to the DTCs, both with key on, engine off (KOEO) and key on, engine running (KOER.) You may observe more valuable clues in the data stream to speed up your diagnosis. For example, let's say that you observe a barometric pressure (BARO) reading of 22.5 in/Hg. This is a reading that is much lower than expected, because barometric pressure in most areas never drops below 28.5 in/Hg, even in a severe storm. You now have a direct line of approach to a condition that may cause the engine to run very poorly. While it is not advisable to replace the sensor without component testing to verify the problem, it may become the prime suspect. This would be especially true if the same sensor is used to measure both BARO and MAP, and the MAP reading is normal with KOER.

Step Seven: Perform Specific DTC Routines

Now is the time to look up the specified test routines for the DTCs you have recorded. Carefully following the prescribed testing routine should lead you straight to the root cause.

A few rules for performing the specific diagnostic tests:

1. When more than one type of failure is indicated, check and repair sensor failures first. The sensor inputs can cause other, interrelated DTCs to be set.

2. Use the exact diagnostic method recommended in the manufacturer's test procedure. Take care to meet all of the steps in the testing process and follow the routine exactly.

3. If you are testing a component rather than a circuit, be sure to check the specifications for the measurement. In general, you will test components with an ohmmeter while they are disconnected from their circuits.

4. If your component tests normally but you still suspect it is defective, remove it for a careful visual inspection. It is possible for a component to test electrically normal, yet have a defect that only shows up in actual use.

If you have completed the manufacturer's DTC diagnostic routine and have still not arrived at a definite conclusion, go back to basics one more time. Test the individual components and connections again, including a careful visual inspection. Some disassembly may be required in this procedure.

This situation is likely to occur with misfire DTCs. For example, on an engine with an individual ignition coil for each cylinder, a coil may burn through some of its insulation internally, but show no external sign of failure. Testing the resistance of the primary and secondary circuits of the coil with an ohmmeter may not indicate a problem. Yet, the cylinder continues to misfire, even with a new spark plug installed. Figure 5–6 shows two defective ignition coils with normal resistance tests. The "back to basics" approach in this case is to swap coils with another cylinder and see if the misfire DTC changes with the coil. If it does, you know you have an internal problem with the coil. If not, the problem has to be somewhere in the circuit that is responsible for firing the coil on that misfiring cylinder. Start at the connector to the coil and work your way back to the PCM and the power source.

Figure 5–6   Two bad coils. Both coils tested okay electrically, but both were causing severe misfire. The coil on the left has a burned area on the insulations but the right-hand coil looks just fine externally.

In the next chapter, we will examine specific scan tool diagnostic routines, as well as testing methods for circuits and components in greater detail.

Step Eight: Verify Successful Repair

One of the most common complaints from customers concerning auto repairs is that the work was not done properly the first time. This situation creates distrust of the repair facility and the technician as well. The customer wonders if the repairs he paid for were ever needed in the first place. He or she doubts the skill and honesty of the technician who performed the job. In addition, the customer is further inconvenienced by having to return the vehicle to the shop again and take whatever steps are necessary to be without the vehicle while it is being repaired again.

It is easy to see how exasperating this scenario is for both the customer and the technician. From our previous discussions about enabling criteria and drive cycles, it is also easy to see how a repair cannot be verified by a single, quick road test. To road test the vehicle properly, the technician must be able to produce the conditions that caused the monitor failure and lit the MIL. This step has one unfortunate element, which is the time it takes to perform such a detailed road test, or two, under the specific operating conditions to set the DTC(s). In fact, a global drive cycle can take 30 minutes or more. However, failure to take the time to perform this step makes the success of the repairs a gamble. The decision to take extra time in the interest of maintaining good customer relations and reputation is a tough one, especially if you are almost certain that you found and fixed the problem and/or there are more cars on the lot waiting to be repaired.

Review and Reinforcement Questions

1. A vehicle has a wide variety of stored DTCs, including several emission control component subsystems and the transmission. Technician A says to replace the PCM. Technician B says to study the wiring diagrams and try to find a common circuit that all affected components share. Who is right?

   a. A only

   b. B only

   c. Both A and B

   d. Neither A nor B.

2. What is most likely indicated by a 12-volt battery OCV reading of 10.4 volts?

   a. Sulphated plates

   b. A "dead" cell

c.   A low state of charge

d.   A dirty connection

3.   When performing a voltage drop test on a 12-volt circuit segment, the DMM reads 3.1 volts. This means:

a.   The circuit is operating at only 3.1 volts

b.   The circuit has high resistance

c.   The circuit has a bad ground at the PCM

d.   The circuit is partially shorted

4.   An engine with individual ignition coils for each cylinder has an intermittent misfire DTC set for cylinder 1. Technician A says to swap the coil for cylinder 1 with another cylinder and see if the misfire moves. Technician B says to remove the spark plug and check its resistance. Who is right?

a.   A only

b.   B only

c.   Both A and B

d.   Neither A nor B

5.   A torque converter clutch will not disengage when braking, but operates normally otherwise. Technician A says the torque converter clutch is binding and the torque converter must be replaced. Technician B says the brake on/off (BOO) switch may be causing the problem. Who is right?

a.   A only

b.   B only

c.   Both A and B

d.   Neither A nor B

6.   The display of information seen on a scan tool is a second-generation translation (a translation of a translation). True or False?

a.   True

b.   False

6

Performing Specific Manufacturer's Diagnostic Routines

As we have mentioned before, OBD II is standardized in many ways, but in some ways it is not. Although the definition of the DTCs and the information required to be available are universal, the specific diagnostic routines recommended for a given DTC or driveability system often vary considerably from one year, make, and model to another. Even the same vehicle model, equipped with two different engines or two different combinations of options, may have substantially different diagnostic routines for the same DTC, as well as different drive cycles for their monitors.

In this chapter, we are going to look at several actual diagnostic routines as recommended by the respective manufacturers. The routines are not copied verbatim, but the steps and sequences are accurate. In the interest of clarity, some text or steps have been eliminated or renumbered. The use of additional reference material may be called for in the actual routine, such as PCM pin numbers, wire/terminal identification of sensor and harness connections, and component locations. After each routine, you will find the author's troubleshooting tips and tricks to help you when diagnosing actual vehicles.

Same DTC, Different Routines

All three of the following routines are for the same DTC, but for different vehicle manufacturers. Comparisons of these three diagnostic routines and tests point out the differences in "similar" routines as described above; the routines also illustrate typical step-by-step testing that encompasses the use of the test light, DMM, and scan tool. All examples are for the following DTC and conditions:

P0135, Oxygen Sensor Heater Circuit, Sensor 1, Bank 1 is set. MIL is on continuously. The scan tool would indicate the description as "Heater Circuit, HO$_2$S1B1."

Example One: 2001 Dodge Stratus SE, 2.5L 4 Cylinder Engine

Note The manufacturer in this case does not specify making basic system checks and visual inspections. This step was not overlooked. The scan tool referred to in this procedure is the manufacturer's proprietary diagnostic unit. Some of the functions shown may not be available with every scan tool.

Testing

1. Using scan tool, retrieve DTCs. If scan tool display "Global Good Trip" counter is zero, go to next step. If voltage is 0.4 to 0.6 volt, go to step 2. If voltage is not 0.4 to 0.6 volt, go to step 3.

2. Turn ignition on (KOEO). Using scan tool, perform O$_2$ sensor heater test. Observe HO$_2$S1B1 signal voltage. Wait 2 minutes for voltage to stabilize. If voltage is 0.4 to 0.6 volt, go to next step. If voltage is not 0.4 to 0.6 volt, go to step 4.

3. At this time, problem does not exist or is intermittent. Using scan tool, read Freeze Frame data. Attempt to operate vehicle in similar conditions. Pay particular attention to code set conditions—RPM, ECT, MAF, TPS, and MAP. If DTC reoccurs, go to next step. If DTC does not reoccur, test is complete.

4. Turn ignition off. Disconnect the sensor harness connector. HO$_2$S1B1 is located upstream of the catalytic converter. Inspect both sides of connector for corroded, damaged, pushed out, or miswired terminals. If connector is okay, go to next step. If connector is not okay, repair or replace as necessary.

5. Turn ignition on. Using scan tool, perform O_2 sensor heater test. Measure voltage between ground and terminal number 2 (Dark Green/Orange wire) at O_2 sensor harness connector. If voltage is 10 volts or more, go to next step. If voltage is less than 10 volts, repair open circuit in Dark Green/Orange wire between O_2 sensor harness connector and Automatic Shutdown (ASD) relay.

6. Turn ignition off. Measure resistance between terminal numbers 1 and 2 at O_2 sensor connector, component side. If resistance is 4 to 7 ohms, go to next step. If resistance is not 4 to 7 ohms, replace sensor.

7. Measure resistance between ground and terminal number 1 (Black wire) at O_2 sensor harness connector. If resistance is less than 5 ohms, replace sensor. If resistance is 5 ohms or greater, repair open circuit in Black wire between O_2 sensor harness connector and ground. Ground point is located on driver's side strut tower.

Tips and Tricks   Note that this procedure has a "dead end" at step 3. If the condition is not present and cannot be reproduced, the DTC can be erased with a "no problem found" conclusion. It may occur to you that an intermittent problem such as this might be discovered by performing the visual checks in step 4, as well as the preliminary basic checks and inspections that the manufacturer did not even call for. Another pitfall occurs in step 5. The possibility of an open circuit to the ASD relay is raised, but checking the terminal connections at the relay and checking for resistance in the relay itself are not mentioned.

If you spotted either of these pitfalls, congratulations! You are thinking like an experienced driveability technician. It is a helpful habit to read the entire diagnostic routine carefully before you start testing. Look for pitfalls like those just discussed.

Now, let's look at another example of a P0135 diagnostic routine.

Example Two: 1999 Chevrolet Camaro, 3.4L V-6 Engine

Note   This manufacturer *does* call for running a series of basic checks and inspections, similar to those described in the previous chapter. Once the basic checks are complete and satisfactory, the next step is to perform the step-by-step procedure. This test begins with "allowing the engine to completely cool down before proceeding." However, no definition is provided for what temperature the engine should be allowed to reach during the cooling process. Like the previous example, the scan tool referred to is the manufacturer's proprietary unit. Therefore, some functions may not be available with every scan tool.

While reading this procedure, compare the steps to the previous routine.

Testing

1. Once the engine has cooled, turn ignition on (KOEO). Using scan tool, monitor HO_2S1B1 signal voltage. If voltage changes from about 0.450 volt to greater than 0.600 volt or from about 0.450 volt to less than 0.300 volt, refer to "Diagnostic Aids" (listed below at end of procedure). If voltage does not change as indicated, go to next step.

2. Remove and inspect fuse for HO_2S1B1 ignition feed circuit. If fuse is blown, go to step 15. If fuse is okay, go to next step.

3. Turn ignition off. Disconnect HO_2S1B1 harness connector. With test light connected to ground, probe ignition feed circuit terminal at sensor harness connector. If test light illuminates, go to next step. If test light does not illuminate, go to step 7.

4. Connect a test light between heater ground circuit and ignition feed circuit terminals at HO_2S1B1 harness connector. If test light illuminates, go to next step. If test light does not illuminate, go to step 8.

5. Allow sensor to cool for a minimum of 10 minutes. Using DMM, measure resistance between ignition feed circuit and heater ground circuit terminals at HO_2S1B1 pigtail (component side). If resistance is 3 to 10 ohms, go to step 9. If resistance is not 3 to 10 ohms, go to step 14.

6. Repair open circuit in HO_2S1B1 ignition feed circuit. After repairs, go to step 15.

7. Repair open circuit in HO_2S1B1 heater ground circuit. After repairs, go to step 15.

8. Check for poor connection(s) at HO_2S1B1 harness connector terminals. Repair as necessary. After repairs, go to step 15. If terminals are okay, go to next step.

9. Turn ignition off. Disconnect PCM harness connectors. Measure resistance of signal circuit and ground circuit between PCM and HO_2S1B1 harness connector. If both readings are less than 5 ohms, go to next step. If one or both readings are 5 ohms or greater, repair "open" in appropriate circuit. After repairs, go to step 15.

10. Check for poor signal circuit or ground circuit terminal connection at HO_2S1B1 harness connector. Repair as necessary. After repairs, go to step 15. If terminals are okay, go to next step.

11. Check for poor HO_2S1B1 ground circuit terminal at PCM. Repair as necessary. After repairs, go to step 15. If terminal is okay, go to next step.

12. Check for poor HO_2S1B1 signal circuit terminal at PCM. Repair as necessary. After repairs, go to step 15. If terminal is okay, go to next step.

13. Replace HO_2S1B1. After repairs, go to step 15.

14. Locate and repair short to ground in HO_2S1B1 ignition feed circuit. Replace fuse. After repairs, go to next step.

15. Allow engine to completely cool down before proceeding. Using scan tool, clear DTCs. Turn ignition on (KOEO.) Using scan tool, monitor HO$_2$S1B1 voltage. If voltage changes form about 0.450 volt to greater than 0.600 volt or from about 0.450 volt to less than 0.300 volt, system is okay. If voltage does not change as indicated, go to step 2.

Diagnostic Aids Check for poor connection at PCM. Check for damaged wiring harness. An intermittent problem can be caused by a poor connection, rubbed-through wire insulation, or broken wire inside insulation. Retrieving *failure records* vehicle mileage since diagnostic monitor last failed may help determine how often condition that caused DTC to set occurs.

Tips and Tricks Note that the okay ranges for the oxygen sensor heater on this vehicle are similar to those in the section "Example One: 2001 Dodge Stratus SE, 2.5L 4 Cylinder Engine," but not exactly the same. You must always be sure that you know the normal measurements and operating characteristics of a component you are testing. This is especially true if you are dealing with an intermittent problem or testing components to find the cause of a driveabiltiy symptom where no DTC is set.

Another Tip When reading the DTCs for the first time, take a look at the Failure Records. If the failure has not occurred very recently, it is less likely to be caused by a defective sensor. If this is the case, pay closer attention to the wiring inspections and tests for power and grounds. You may want to use your DMM and check the resistance in the ground circuit rather than relying on a test light.

A timesaver in this routine would be welcome, since there are three different steps where allowing the engine or the oxygen sensor to cool down is called for. One way to avoid at least one of these cool-down cycles is to perform step 5 before step 1. Since the engine had to cool down before beginning the diagnostic routine, the oxygen sensors also would cool provided that the ignition was left "Off." Why not unplug the suspect sensor while it already is cool and check the resistance of the heater before going through the steps of testing the circuits? While the harness connector is unplugged, inspect the connector terminals for corrosion or other damage. If desired the test light checks for the ignition feed and ground circuits, steps 3 and 4, also could be performed while the harness connector is open.

Let's switch vehicles again and check the same DTC, P0135, one more time.

Example Three: 2000 Ford Taurus SE, 3.0L V-6 Engine

> **Note** This manufacturer structures their diagnostic reference information and routines differently from the two previous examples. The DTCs are first displayed in chart form, where the technician is directed to a particular point within a certain diagnostic test. Each of these tests is comprehensive in nature, being directed at an entire system.

Since each test is designed to diagnose more than one DTC or component, there may be several diagnostic routines within a single system test. The different DTC diagnostic routines will begin at different points within the test. To make things more interesting, the steps in the diagnostic routines often are not numbered sequentially! In the interest of clarity, the routines within the system test that are not related to DTC P0135 are not shown. Once again, the scan tool assumed to be in use is the manufacturer's proprietary unit. Therefore, some functions may not be available with every scan tool. The chart listing for DTC P0135 on this model calls for "Test H: Fuel System" and for testing to begin at step 1.

Testing

1. DTC P0135, P0141, P0155, or P0161: HO_2S Heater Circuit.

   DTCs received separately indicate a short to ground or open circuit in HO_2S heater circuit.

   DTCs received in pairs, such as P0135 and P0155 or P0141 and P0161, indicate HO_2S heater circuit is shorted to a power source of more than 2 volts. HO_2S DTC identification is as follows:

   - DTC P0135 is for HO_2S 1/1 (Bank 1, Sensor 1)
   - DTC P0141 is for HO_2S 1/2 (Bank 1, Sensor 2)
   - DTC P0155 is for HO_2S 2/1 (Bank 2, Sensor 1)
   - DTC P0161 is for HO_2S 2/2 (Bank 2, Sensor 2)

   Possible causes are:

   - Signal shorted in wiring harness or HO_2S
   - Water in connectors
   - Cut or pulled wires

- Open in GND (Ground) or VPWR circuit (called Ignition Feed in previous procedure)
- Corroded terminals
- Faulty HO$_2$S heater

Inspect HO$_2$S connectors for loose, damaged, or corroded terminals. Repair as necessary. If HO$_2$S connectors are okay, go to next step.

2. Perform KOEO self-test.

   Start engine and operate at 2000 RPM for one minute. Turn ignition switch to Off position. Perform KOEO On-Demand Self-Test. ("On-Demand" is this man-ufacturer's description for problems currently present.) If DTC P0135, P0141, P0155, or P0161 is present, go to next step. If specified DTCs are not present, fault is intermittent. Go to Test Z (see section "Intermittent Diagnostic Routine Example, 2000 Ford Taurus SE, 3.0 L V-6 Engine" later in this chapter).

3. Check for VPWR voltage at HO$_2$S heater wiring harness connector.

   Turn ignition switch to Off position. Disconnect suspect HO$_2$S. Inspect wiring harness for damage and repair as necessary. Turn ignition switch to On posi-tion. Measure voltage between SIG RTN (Signal Return) and VPWR (Ignition Feed) terminals at suspect HO$_2$S harness connector. If voltage is more than 10.5 volts, go to next step. If voltage is 10.5 volts or less, check circuit fuse. Replace as necessary. If fuse is okay, repair open circuit.

4. Check for Shorted Circuit.

   Disconnect scan tool from DLC. Disconnect PCM connector(s). Inspect con-nector(s) for loose, damaged, or corroded terminals. Repair as necessary. Measure voltage between PCM harness connector pin for suspect HO$_2$S HTR (suspect oxygen sensor heater) circuit terminal and PWR GND (power ground) terminal. Also, measure resistance between HO$_2$S HTR terminal and SIG RTN and VPWR terminals at suspect HO$_2$S harness connector. If both resistance readings are more than 10,000 ohms, go to next step. If any resistance read-ing is 10,000 ohms or less, repair short circuit.

5. Check for Open Circuit.

   Leave ignition switch in Off position and suspect HO$_2$S disconnected. Measure resistance between HO$_2$S HTR terminal at suspect HO$_2$S harness connector and appropriate HO$_2$S HTR terminal at PCM connector. If resistance is 4 ohms or more, repair heater circuit for open or excessive resistance in wiring harness. If resistance is less than 4 ohms, go to next step.

6. Check HO$_2$S Heater Resistance.

   Turn ignition switch to Off position. With suspect sensor disconnected, meas-ure resistance between HO$_2$S HTR terminal and VPWR terminal at suspect HO$_2$S. If resistance is 3 to 30 ohms, go to next step. If resistance is not 3 to 30 ohms, replace HO$_2$S.

7. Check HO$_2$S Case for Short to VPWR, HTR or SIG RTN.

   Measure resistance between HO$_2$S component case (metal housing of sensor) and HO$_2$S HTR, SIG RTN, and VPWR terminals. If all resistance measurements are more than 10,000 ohms, replace PCM and program replacement unit. If any resistance is 10,000 ohms or less, replace HO$_2$S.

Note   This manufacturer calls for verification of repair success by clearing DTCs and repeating the quick test as described above. No recommendation is provided for verification using a drive cycle or "trip" procedure. However, drive cycle information is available for this vehicle.

Tips and Tricks   In this diagnostic routine, the manufacturer calls for a visual inspection of the connector terminals as part of the first step. However, testing the resistance of the oxygen sensor heater circuit is not called for until after testing for short and open circuits, which involves at least partial disassembly of the PCM. It may be more efficient to test the heater circuit resistance in the first step, while the connector is open and the sensor is cool.

This manufacturer has an interesting approach to assessing multiple DTCs, which can be beneficial to remember before beginning hands-on diagnostic tests on any vehicle. In step 1 the DTCs are treated differently if they are received in certain pairs than if one individual DTC is set. If any of the listed pairs of DTCs are set, a definite cause is cited and the diagnosis does not require a step-by-step routine. This is because both oxygen sensors in the pair of DTCs share only a portion of the heater circuits.

The concept of looking for shared circuits can often lead the technician to a single repair for multiple DTCs. When more than one DTC are set, it is a good idea to check for any portion of wiring, connectors, ground connections, or power supply that are shared by some or all of the components related to the DTCs. Going straight to this point and performing a quick visual inspection may save the trouble of performing the entire diagnostic routine. This technique is especially helpful if the DTCs are the result of intermittent problems.

There are two ways to assess the shared circuit components. You can study the vehicle wiring diagrams to look for any areas where the circuits come together, or you can study the individual DTC diagnostic routines and look for testing steps that are the same for each. Studying the wiring diagrams is preferable, because it is possible that the *common link* is a component or connector that is not included in any of the individual DTC testing steps. Look very carefully when DTCs are seemingly unrelated.

EVAP System Example, P0442, Small Leak Detected, 1999 Chevrolet Suburban C1500, 5.7L V-8 Engine

This DTC indicates that OBD II has detected a small leak during its monitoring of the EVAP system. As you read the testing procedure, you will find that the causes for such a fault can be either very basic or rather well-hidden. You also will see that other interrelated DTCs can cause false input data to be sent to the PCM.

The diagnostic routine assumes that an EVAP system pressurizing tester and ultrasonic leak detector are available, as well as the manufacturer's proprietary scan tool. Without this equipment, some of the listed test steps may not be possible. Note that pressure and vacuum readings in the EVAP system are measured in *inches of water*, (in/H_2O), rather than the customary units of measurement. We will join the test procedure after the basic tests and inspections have already been performed and satisfactorily passed.

Testing

1. Check if DTC P0446, P0452, or P0453 is set. If any of these DTCs are present, diagnose those DTCs first. If none of these DTCs are set, go to next step.

2. Turn ignition off. Remove fuel filler cap. Turn ignition on. Using scan tool, read fuel tank vacuum. If scan tool displays 0 in/H_2O, go to next step. If scan tool does not display 0 in/H_2O, repair faulty fuel tank sensor circuit.

3. Zero both pressure and vacuum gauges on EVAP Pressure/Purge Diagnostic Station. Reinstall fuel filler cap. Read and record Failure Records data for this DTC. Clear DTCs. Using scan tool, command EVAP vent valve on (closed). Connect EVAP Pressure/Purge Diagnostic Station to EVAP service port. Using diagnostic station, pressurize EVAP system to 15 in/H_2O. Monitor pressure on diagnostic station gauge. If pressure decreases to less than 10 in/H_2O, go to next step. If pressure does not decrease to less than 10 in/H_2O, see "Diagnostic Aids" (below, at end of routine).

4. Disconnect fuel tank vapor and EVAP purge lines from EVAP canister. Plug fuel tank vapor line fitting at EVAP canister. Connect a hand vacuum pump to EVAP purge line fitting at EVAP canister. Ensure EVAP vent solenoid is still commanded on (closed). Attempt to apply 5 in/Hg vacuum. If vacuum can be obtained and held, go to step 10. If vacuum cannot be maintained, go to next step.

5. Leave system connected as in previous step. Disconnect and plug vent hose at vent valve. Attempt to apply 5 in/Hg vacuum with hand pump. If vacuum can be obtained and held, go to step 7. If vacuum cannot be maintained, go to next step.

6. Check vent hose for leak. Repair as necessary. After repairs, go to step 11. If hose is okay, go to step 8.

7. Replace EVAP vent valve. After replacing vent valve, go to step 11.

8. Replace EVAP canister. After replacing canister, go to step 11.

9. Check for missing or faulty fuel cap. Check for disconnected or leaking fuel tank vapor line. Check for disconnected or damaged EVAP purge line. Repair as necessary. After repairs, go to step 11. If no problem is found, go to next step.

10. Using scan tool, command EVAP vent solenoid on (closed). With EVAP Pressure/Purge Diagnostic Station connected to EVAP service port, attempt to pressurize EVAP system to 15 in/H_2O by leaving EVAP pressure/purge diagnostic station control knob in Pressurize position. Using ultrasonic leak detector, locate and repair EVAP system leak. It may be necessary to lower fuel tank to check connections on top of tank. After repairs, go to next step.

11. Turn ignition on, engine off. Using scan tool, command EVAP vent solenoid on (closed). With EVAP Pressure/Purge Diagnostic Station to EVAP service port, pressurize EVAP system to 15 in/H_2O. Monitor pressure on diagnostic station gauge. Turn the diagnostic station rotary switch to Hold position. If pressure decreases to less than 10 in/H_2O within 2 minutes, return to step 1. If pressure does not decrease as specified, go to next step.

12. Using scan tool, clear DTC. Using scan tool, perform Service Bay Test for EVAP system. After performing test, go to next step.

13. Using scan tool, select Read and Record Info, Review Info function. If any undiagnosed DTCs are displayed, diagnose affected DTCs. If no DTCs are displayed, system is okay.

Tips and Tricks   The first tip is a precaution: *Never* pressurize the EVAP system with compressed air as a substitute for use of the proper equipment in any diagnostic routine! Even if the amount of pressure is properly controlled, the oxygen present in compressed air is hazardous in the presence of gasoline vapors contained in the fuel tank and other system components. Equipment designed to pressurize the EVAP system is specially designed for safety and introduces noncombustible gas into the system instead of compressed air.

The order of testing in this diagnostic routine is quite logical and efficient. This system is easy to gain access in some places, but quite difficult in others. The diagnostic routine tells us to check the components and access the system for pressure testing at the easy end first, the area of the EVAP canister. If no problem is found there, the technician is directed to the parts of the system that are more difficult to access, the under-car lines and the fuel tank itself. The only opportunity to improve upon this routine would be to have a close look at the fuel filler cap before going to the trouble of checking the fuel tank sensor, lines, and vent solenoid.

If an obvious defect such as a broken o-ring seal or other physical damage is seen, or if the cap appears to be an aftermarket replacement, it might be more efficient to pressurize the system and look for a leak at the cap first, rather than waiting until step 9.

Engine Misfire Example, P0302, Misfire on Cyl. No. 2, 2002 Chevrolet Cavalier, 2.4 L 4 Cyl. Engine

The diagnostic routine is the same for any cylinder-specific misfire DTC. It may be necessary to refer to the vehicle wiring diagrams, component location charts, and connector identifications to locate proper test points.

On this vehicle, the PCM uses information from the Ignition Control Module and the Camshaft Position Sensor (CMP) in order to determine when an engine misfire is occurring. By monitoring the momentary variations in the crankshaft RPM for each cylinder, the PCM is able to detect individual misfire events. A misfire rate that is high enough can cause the three-way catalytic converter to overheat under certain conditions. The MIL will flash on and off when the conditions for three-way catalytic converter damage are present.

Testing

1. Perform the diagnostic system check -engine controls. After performing basic tests and inspections with satisfactory "passed" results, go to next step.

2. Turn ignition On with engine Off. Using scan tool, check to see if any injector DTCs are set. If yes, follow diagnostic routine(s) for those DTCs first. If no, go to next step.

3. Turn ignition Off. Perform a visual and physical inspection. Make any necessary repairs. Were any repairs necessary? If yes, go to step 25. If no, go to next step.

4. Start and idle engine. Observe the Misfire Current Counters with the scan tool. Is the Misfire Current Counter for the affected cylinder incrementing? If yes, go to next step. If no, go to step 6.

5. Turn ignition Off. Remove air cleaner outlet resonator assembly. Disconnect all fuel injector electrical connectors. Disconnect the IC Module 11-pin electrical connector. Temporarily remove the ignition coil and the electronic IC Module assembly. Reconnect the 11-pin electrical connector. Install an ignition system diagnostic jumper harness into spark plug boot assemblies. Install a spark tester on the spark plug jumper wire for the affected cylinder spark plug. Remove spark plug boot assembly from companion cylinder of ignition coil housing. Install a jumper wire from the companion spark plug connector of ignition coil housing to ground. Connect a jumper wire from the IC Module

assembly to ground. Crank engine with remaining spark plug wires still connected. Does spark tester spark? If yes, go to step 7. If no, go to step 10.

6. Turn ignition On, with engine Off. Review the Freeze Frame data and record the parameters. Start the engine. Operate the vehicle within the Freeze Frame conditions and enabling conditions for the DTC. Observe the Misfire Current Counters using the scan tool. Is the Misfire Current Counter for the affected cylinder incrementing? If yes, return to step 5. If no, go to step 25.

7. Turn ignition Off. Remove the spark plug from the affected cylinder. Exchange the spark plug with a spark plug from a known good cylinder. Reconnect all the fuel injector electrical connectors. Install the spark plug boot assembly to the affected cylinder spark plug connection at the ignition coil housing. Reconnect the spark plug jumper wire to the spark plug boot assembly. Start the engine. Run the vehicle in the same conditions as when the misfire was present. Is the Misfire Current Counter for the affected cylinder incrementing? If yes, go to next step. If no, go to step 9.

8. Check for leaking fuel injector. Refer to diagnostic routines for DTC P0201-P0204, starting at step 5. If the fuel injector is okay, check for a basic engine problem affecting compression or valve operation. Did you find and correct the condition? If yes, go to step 25. If no, go to the next step.

Author's Note Step 8 is the end of the basic diagnostic routine. The remaining steps are corrective subroutines to be used when problems are indicated in the course of steps 1 to 7.

9. Turn ignition Off. Replace the defective spark plug. After repairs, go to step 25.

10. Turn ignition Off. Remove the spark plug boot assemblies from the affected coil. Disconnect the IC Module 11-pin connector. Connect a DMM between the secondary ignition coil terminals at the ignition coil housing. Is the resistance between 4000 and 8000 ohms? If yes, go to next step. If no, go to step 12.

11. Remove the ignition coil housing. Disconnect the coil harness electrical connector from the IC Module. Reconnect the 11-pin IC Module electrical connector. Connect a test light to battery positive (B+). Probe the affected cylinder coil control terminal with the test light. Crank the engine while watching the test light. Does the test light blink? If yes, go to step 13. If no, go to step 14.

12. Remove the affected cylinder ignition coil from the ignition coil housing. Recheck the resistance between the secondary terminals of the affected cylinder ignition coil. Is the resistance between 4000 and 8000 ohms? If yes, go to step 15. If no, go to step 16.

13. Turn ignition Off. Remove both ignition coils from the ignition coil housing. In this step, handle coil connectors carefully to avoid creating a short circuit

and blowing a fuse, which could lead to misdiagnosis. Disconnect electrical connectors from ignition coils. Reconnect electrical connector to IC Module. Connect a test light to ground. Turn ignition On with engine Off. Probe both ignition coil feed terminals "B" with test light. Does the test light illuminate for both circuits? If yes, go to step 19. If no, go to step 20.

14. Turn ignition Off. Disconnect IC Module 11-pin connector. Visually and physically inspect IC Module and electrical connector to see if connections are clean and tight. Make any necessary repairs. Did you find and correct the condition? If yes, go to step 25. If no, go to step 17.

15. Check and replace ignition coil spring for affected coils or ignition coil housing as necessary. When repairs are complete, go to step 25.

16. Replace the ignition coil for the affected cylinders. When repair is complete, go to step 25.

17. Reconnect IC Module 11-pin electrical connector. Disconnect PCM electrical connector C1. Reconnect affected cylinder spark plug boot assembly to spark plug connector at ignition coil housing. Reinstall the ignition system diagnostic jumper harness into spark plug boot assemblies. Install a spark tester on the spark plug jumper wire for the affected cylinder spark plug. Remove spark plug boot assembly from companion cylinder of ignition coil housing. Install a jumper wire from the companion spark plug connector of ignition coil housing to ground. Connect a jumper wire from the IC Module assembly to ground. Turn ignition On with engine Off. Connect a test light to battery positive (B+). Do not leave the test light connected to the PCM IC circuit connector for longer than 5 seconds at a time, or damage to the IC Module or ignition coil. Momentarily touch the test light to the IC timing control circuit terminal for the affected cylinder at the PCM C1 electrical connector, harness side. Each time the test light is removed, a spark should occur at the spark tester. Is there a spark present when you remove the test light from the IC timing control terminal? If yes, go to step 23. If no, go to next step.

18. Turn ignition Off. Check for the following conditions in the IC timing control circuit:

   ● An open circuit

   ● A short circuit to ground

   ● A short circuit to B+

   ● A poor electrical connection at IC Module

   Repair the circuit as necessary. Were any repairs necessary? If yes, go to step 25. If no, go to step 22.

19. Check the affected cylinder ignition coil control circuit for the following conditions:

   ● An open circuit

   ● A short circuit to ground

- A short circuit to B+
- A poor electrical connection at IC Module and/or ignition coil

Repair the circuit as necessary. Were any repairs necessary? If yes, go to step 25. If no, go to step 21.

20. Turn ignition Off. Repair open circuit in the ignition positive voltage circuit between the No. 1-4 and the No. 2-3 ignition coils electrical connectors. After repairs, got to step 25.

21. Turn ignition Off. Check ignition positive voltage circuit for a poor electrical connection at the ignition coil of the affected cylinder. If electrical connection is okay, replace spark plug boot assembly. After repairs, go to step 25.

22. Turn ignition Off. Replace the IC Module. When repair is complete, go to step 25.

23. Turn ignition Off. Inspect the PCM electrical connector and connections. Repair the electrical connector and connections as necessary. Did you find and correct the condition? If yes, go to step 25. If no, go to next step.

24. Replace PCM. Perform PCM relearn procedure, including reprogramming and the CKP System Variation Learn Procedure. After replacing PCM, go to next step.

25. Reconnect all previously disconnected components, if not already reconnected. Install the air cleaner outlet resonator assembly. Clear the DTCs with scan tool. Start the engine. Idle the engine at normal operating temperature. Operate the vehicle within the conditions for running this DTC and also for running P0420. Does the scan tool indicate that these diagnostics have run and passed? If yes, go to next step. If no, go to step 2.

26. Turn ignition On, with engine Off. Using scan tool, observe the stored information and capture info. Does the scan tool display any DTCs that you have not diagnosed? If yes, see DTC definitions and specific diagnostic routine for each of those DTCs. If no, system is okay.

Tips and Tricks   This diagnostic routine encompasses a wide variety of troubleshooting techniques that you will find useful in many situations you may encounter. The test light was used in an unusual fashion to bypass the PCM and command the IC Module to fire the ignition coil. Since the ignition coils are mounted under a trim plate and then inside of a housing as well, a set of spark plug jumper wires had to be installed so the engine could be operated with the coils separated from the spark plugs. The technician also was directed to stop troubleshooting electronically and perform inspections *visually and physically* at various stages of the diagnosis. This is good advice, which encompasses carefully looking at the suspect component, circuit, or connection and wiggling the wiring and connector to see that all connections are clean and tight.

When disconnecting and reconnecting components in the course of a routine such as this example, you may inadvertently "fix" a loose, dirty, or corroded terminal connection. This is one reason that a problem may disappear in the middle of a diagnostic routine. If this happens, carefully check the last connections that were disturbed.

Sometimes, a visual inspection may be the only hard evidence you will find in identifying the root cause of the problem. The ignition coil is a classic example of a part that can develop a mechanical failure, such as a crack in its insulation, which will cause it to misfire. Yet, all electrical tests performed with the DMM may show normal values. In addition, the misfire may not occur at idle, but only under certain driving conditions. Visual inspection may reveal the problem, or the coil may have to be swapped with the coil for another unaffected cylinder to prove the diagnosis. If the suspect coil is defective, the Misfire DTC will change to the cylinder where it was moved.

In step 24, PCM replacement and reprogramming is called for. Reprogramming the replacement PCM with the latest software version for the specific vehicle in which it is installed customarily requires the use of the manufacturer's proprietary equipment and service information. However, aftermarket sources, such as major parts suppliers, do have some limited reprogramming capabilities as we go to press. Some reprogramming may be accomplished before installing the PCM in the vehicle, but other programming procedures must be performed after everything is installed and reconnected. We will further discuss reprogramming in the next section, "Component Testing."

Intermittent Diagnostic Routine Example, 2000 Ford Taurus SE, 3.0 L V-6 Engine

The following routine is recommended by Ford Motor Company in cases where the DTC or symptom appears to be the result of an intermittent condition. The technician may be directed to this test, identified as Test Z, within one of the steps of another diagnostic routine. In the section "Example Three: 2000 Ford Taurus SE, 3.0L V-6 Engine," the direction to Test Z was given if the DTC did not reset during the KOEO test performed in step 30. The Intermittent Test is intended to diagnose and isolate intermittent concerns for:

- All engine control subsystems
- Vehicles with coil pack ignition systems

Testing

1. **Direction for Intermittent Diagnostic Path.**

   The technician must gather information concerning symptoms, DTCs, Freeze Frame data from the customer and the scan tool, then decide what parameters are suspected, or related to the intermittent problem. Based on the conclusions reached, the testing will utilize parameter identifications (PIDs), which may be tested with the scan tool, ignition tester, and/or DMM. The normal sensor ranges and other individual parameter measurement values must be known before testing begins.

2. **Perform PCM reset procedure.**

   Ensure Freeze Frame data has been recorded before resetting PCM.

3. **Select PIDs related to symptom.**

   Scan tool and a list of PIDs must be used to indicate area of fault. Use a Reference Value Symptom and Reference Value/PID Measurement Signal table in appropriate Pin Voltage/PID charts. Note and make a list of PIDs related to symptoms.

4. **Symptom verification** is recommended if:
   - Vehicle is in for a repeat repair
   - No DTCs are present
   - Customer has difficulty describing symptom

   If symptom needs to be verified, go to step 5. If symptom does not need to be verified, go to step 11.

5. **Note all available data to aid on symptom verification.**

   Collect as much data as possible to aid in determining intermittent fault area. Note any Freeze Frame data that was recorded during General Diagnostic and Troubleshooting checks or in step 2. Note any Continuous Memory DTCS (intermittent failures) that were set prior to performing PCM Reset in step 2. Obtain any information from customer that would aid in determining fault area. Go to step 6.

6. **Attempt to recreate symptoms.**

> **Note**   A road test may be required to recreate similar conditions that originally caused DTC to set.

Using scan tool, select and monitor PIDs displayed in Freeze Frame data (if available) and PIDs on list that was developed in step 3. Using Free Frame data recorded earlier, recreate conditions described by each Freeze Frame PID. Pay particular attention to ECT, Load (MAP/MAF), RPM, and VSS. Use any information from the customer to aid in producing conditions for recreating symptoms. When symptom occurs, press trigger on scan tool to begin recording. Refer to scan tool instruction manual for recorder function operation. If symptom is reproduced, go to step 11. If symptom cannot be reproduced, go to next step.

7. **Recreate symptom using intermittent road test procedure** (Steps 7 through 10).

> **Note**   PIDs for PCM outputs found in Pin Voltage/PID Value tables represent commanded values only. Circuit measurements made with DMM indicate actual output status. If DMM circuit measurement differs from scan tool PID value, a circuit fault may be indicated. PCM input PIDs that differ from DMM measurements indicate a possible PCM problem.

Intermittent road test procedure is the last attempt to locate area of problem before physically disturbing vehicle circuits. The road test procedures require monitoring PID values or making circuit measurements with a DMM under four operating conditions—KOEO, Hot Idle, 30 mph road test, and 55 mph road test. Use PID values and circuit values in Pin Voltage/PID tables to compare with actual PID values and circuit measurements during road tests. Turn ignition switch to On position. Using scan tool, select and monitor values. Using a DMM, measure circuits and compare scan tool PID values with DMM readings. If any values are out of range, go to step 11. If all PID values are within range, go to next step.

8. **Recreate symptom using hot idle road test procedure.**

Start engine and allow to idle. Ensure that engine is at normal operating temperature of at least 195°F (87°C). Continue to monitor same PIDs and circuit measurements as in step 7. If any values are out of range, go to step 11. If all values are in range, go to next step.

9. **Recreate symptom using 30 mph road test.**

Drive vehicle at 30 mph and continue to monitor same PIDs and circuit measurements as in step 7. If any values are out of range, go to step 11. If all values are in range, go to next step.

10. **Recreate symptom using 55 mph road test.**

Drive vehicle at 55 mph and continue to monitor same PIDs and circuit measurements as in step 7. If any values are out of range, go to next step. If all values are in range, it is now necessary to physically disturb selected circuit in an attempt to recreate intermittent concern. Go to next step.

11. **Select circuits to be tested from Intermittent Test Table.**

Switch scan tool to PID selection menu. If steps 7 to 10 were performed, select PIDs or circuit values that were out of range or that did not match PID values shown on table or measured with DMM. If steps 7 to 10 were not performed, select PIDs from the list made in step 3. On all applications, go to Intermittent Test Table. Match selected PIDs to corresponding circuit in table. It is possible to have more than one circuit to test. If a PID recording was made with the scan tool in step 6, it may be helpful to replay the recording at this time. Using the table, determine test to perform, input or output. To perform input test, go to next step. To perform output test, go to step 16. If PID or circuit is not listed in Intermittent Test Table, diagnose driveability symptom as a no-code troubleshooting routine. Refer to Troubleshooting—No Codes—EEC-V (Gasoline).

Author's Note   The above reference to troubleshooting routines leads the technician to inspection and individual testing of all components and circuits related to the symptom being repaired. This type of testing is resorted to only if no DTCs were ever set or if the suspect circuits or components to be tested have no listing in the Intermittent Test Table.

12. **KOEO Intermittent Input Test for PCM Sensors.**

Using circuits selected from Intermittent Test Table, select only recommended PIDs to monitor with scan tool. If a PID is not available for circuit, use a DMM to monitor circuit. Go to area of suspect wiring or component fault. Turn ignition to On position. If input is a switch-type component, (A/C request, for example) turn switch on manually. Observe PID or DMM value while lightly tapping on component, and then wiggling sensor wiring harness from component to PCM. If a fault is indicated, PID value or DMM reading will change suddenly. Compare to correct KOEO values listed in appropriate Pin Voltage/PID Value tables. If no fault is indicated, go to next step. If a fault is indicated, isolate fault and repair as necessary. If wiring harness and connectors are okay, replace suspect component(s).

13. **KOER Intermittent Input Test Procedure for PCM Sensors.**

Caution   Ensure hands, clothing, and tools are clear of hot surfaces, cooling fans, drive belts, and other moving components when performing tests with engine running.

Start engine and allow to idle. Continue to monitor PIDs and/or circuits as in step 12. Go to area of suspect wiring or component fault. If input is a switch-type component, turn switch on manually. Observe PID or DMM value while lightly tapping on component, and then wiggling sensor wiring harness from component to PCM. If a fault is indicated, PID value or DMM reading will change suddenly. Compare to correct Hot Idle values listed in appropriate Pin Voltage/PID Value tables. If no fault is indicated, turn ignition switch to Off position and go to next step. If a fault is indicated, isolate fault and repair as necessary. If wiring harness and connectors are okay, replace suspect component(s).

14. **KOEO Intermittent Water Soak Test Procedure for PCM Sensors.**

    Turn ignition switch to On position. Continue to monitor PIDs and/or circuits as in step 13. Go to area of suspect wiring or component fault. If input is a switch-type component, turn switch on manually. Observe PID or DMM value while lightly spraying water on component, circuit, and connectors from component to PCM. Include any relays or relay modules associated with fault. If fault is indicated, value will suddenly change. Compare to correct KOEO values listed in appropriate Pin Voltage/PID Value tables. If no fault is indicated, go to next step. If a fault is indicated, isolate fault and repair as necessary. If wiring harness and connectors are okay, replace suspect component(s).

15. **KOER Intermittent Water Soak Test Procedure for PCM Sensors.**

Caution Ensure hands, clothing, and tools are clear of hot surfaces, cooling fans, drive belts, and other moving components when performing tests with engine running.

Start engine and allow to idle. Continue to monitor PIDs and/or circuits as in step 14. Go to area of suspect wiring or component fault. If input is a switch-type component, turn switch on manually. Observe PID or DMM value while lightly spraying water on component, circuit, and connectors from component to PCM. Include any relays or relay modules associated with fault. If a fault is indicated, PID value or DMM reading will change suddenly. Compare to correct Hot Idle values listed in appropriate Pin Voltage/PID Value tables. If no fault is indicated, turn ignition switch to Off position and go to next step. If a fault is indicated, isolate fault and repair as necessary. If wiring harness and connectors are okay, replace suspect component(s).

16. **KOEO Intermittent Output Test Procedure for PCM Actuators.**

Note PIDs for PCM outputs found in Pin Voltage/PID Value tables represent commanded values only. Circuit measurements made with DMM indicate actual output values. If DMM circuit measurement differs from scan tool PID value, a circuit fault may be indicated.

Using circuits chosen from Intermittent Test Table, select only recommended PIDs to monitor with scan tool. Also use a DMM to compare circuit values with scan tool PID values. If a PID is not available for a circuit, use a DMM to monitor circuit. Turn ignition switch to On position. Using scan tool, access Output Test Mode.

Note   Output Test Mode may not control some outputs, such as fuel injectors or ignition coils. To test outputs not controlled by Output Test Mode, go to step 17.

Caution   Ensure hands, clothing, and tools are clear of cooling fans, as they may start during output tests.

Note   In performing the output tests, remember that the scan tool Output Test Mode shuts off after 10 minutes of operation. Outputs will have to be triggered again to keep them active.

Using scan tool, turn all outputs On. Go to area of suspect wiring or component fault. Observe PID value while lightly tapping on component, and then wiggling sensor wiring harness and connectors from component to PCM. If fault is indicated, value will change suddenly. Also, compare actual values with KOEO values in appropriate Pin Voltage/PID Value table. If no fault is indicated, go to next step. If fault is indicated, isolate fault and repair as necessary. If wiring harness and connectors are okay, replace suspect component(s).

17. **KOER Intermittent Output Test Procedure for PCM Actuators.**

Start engine and allow to idle. Continue to monitor PIDs and/or circuits as in step 16. Go to area of suspect wiring or component fault. Observe PID or DMM value while lightly tapping on component, and then wiggling sensor wiring harness from component to PCM. If a fault is indicated, PID value or DMM reading will change suddenly. Compare to correct Hot Idle values listed in appropriate Pin Voltage/PID Value tables. If no fault is indicated, turn ignition switch to Off position and go to next step. If a fault is indicated, isolate fault and repair as necessary. If wiring harness and connectors are okay, replace suspect component(s).

Note   If an ignition coil mounted on an individual spark plug has been tapped and is suspect, turn ignition switch to Off position. Remove suspect coil. Using DMM, check continuity between spark plug terminal and signal terminal circuit of coil. Observe DMM while lightly tapping on coil. A large fluctuation in resistance indicates an intermittent open circuit in coil. Replace coil as necessary. If coil is okay, check wiring harness.

18. **KOEO Output Intermittent Water Soak Test.**

Warning Avoid contact with PCM, Generic Electronic Module (GEM) or other modules when performing water soak test on electrical components and/or harnesses.

Note Output Test Mode may not control some outputs, such as fuel injectors or ignition coils. To test outputs not controlled by Output Test Mode, go to step 19.

Turn ignition switch to On position. Using scan tool, access Output Test Mode. Turn all outputs On. Continue to monitor PIDs and/or circuits as in step 17. Go to area of suspect wiring or component fault. Observe PID or DMM value while lightly spraying water on component, circuit and connectors from component to PCM. Include any relays or relay modules associated with fault. If fault is indicated, value will suddenly change. Compare to correct KOEO values listed in appropriate Pin Voltage/PID Value tables. If no fault is indicated, go to next step. If a fault is indicated, isolate fault and repair as necessary. If wiring harness and connectors are okay, replace suspect component(s).

19. **KOER Output Intermittent Water Soak Test.**

Start engine and allow to idle. Continue to monitor PIDs and/or circuits as in step 18. Go to area of suspect wiring or component fault. Observe PID or DMM value while lightly spraying water on component, circuit, and connectors from component to PCM. Include any relays or relay modules associated with fault. If a fault is indicated, PID value or DMM reading will change suddenly. Compare to correct Hot Idle values listed in appropriate Pin Voltage/PID Value tables. If no fault is indicated, turn ignition switch to Off position and go to next step. If a fault is indicated, isolate fault and repair as necessary. If wiring harness and connectors are okay, replace suspect component(s).

20. **Check for Intermittent Mechanical Concerns.**

Note An intermittent mechanical problem can cause a good PCM and engine management system to act abnormally.

If not done previously, inspect mechanical systems relating to DTC or symptom. Check the following:

● Check if engine rocks during acceleration. If excessive movement is detected, check motor mounts.

- Check for excessive component movement while driving vehicle during conditions that would cause vibrations (high RPM, rough roads, etc.).
- Check accelerator and transmission linkages for contact or interference.

If any problems are detected, repair as necessary. If no problems are detected, fault cannot be identified at this time. Check for related Technical Service Bulletins (TSBs) for additional help. Diagnostic testing is complete.

> **Tips and Tricks**   The author has seen numerous case studies where the testing called for in the note following step 17 in the previous test revealed no abnormal reading, but the suspect coil was misfiring under driving conditions. Refer to Chapter 5. Remember that ignition coils are among the most likely components to test okay, yet still have a "hidden" fault? If a coil is misfiring with any regularity, it should set a misfire DTC. With an individual coil-on-plug application, the easiest way to verify an intermittent defective coil is to swap the coil with another cylinder and see if the misfire DTC moves with the coil on subsequent road tests.

Back in step 7, the manufacturer stated, "If DMM output circuit measurement differs from scan tool PID value, a circuit fault may be indicated. PCM input PIDs that differ from DMM measurements indicate a possible PCM problem." Well, maybe! It is possible for an incorrect output to be generated directly by the PCM, and the technician is only instructed to test the value at the sensor harness connector. The value must also be tested at the appropriate PCM pin to be sure the problem is in the circuit. On the other hand, inputs can degenerate on the way from the sensor to the PCM for a variety of wiring or connector terminal problems. Again, the value must be measured at the PCM pin, at the other end of the circuit wiring, to verify whether or not it is a PCM fault.

Another possible pitfall appears in step 15. If no fault is indicated in this step, the technician is directed to begin the output tests. In the author's opinion, this path would only be valid if the input being tested was also an output or affected by an output value.

As stated at the beginning of this example, this test appears to be quite complex. However, a number of the steps are repetitive, being similar for input and output testing. This procedure is just another approach to what we have been saying all along. That is, when you encounter an intermittent problem, the most efficient troubleshooting method is often to compare the actual values read to the values the PCM "thinks" it is seeing. In this test, the tool of choice was the DMM. Another efficient method of measurement is the digital storage oscilloscope, or "lab scope," if one is available. We will examine the use of this valuable tool in the next chapter about direct component testing.

Review and Reinforcement Questions

1. The EGR is commanded to open, but no change in STFT occurs. Technician A says the EGR pintle may be stuck shut. Technician B says the EGR passages may be clogged. Who is right?

   a. A only

   b. B only

   c. Both A and B

   d. Neither A nor B

2. An OBD II vehicle has an ECT reading of –22°F on the scan tool. The ECT tests okay with the ohmmeter. Technician A says the ECT circuit is open. Technician B says the ECT circuit is shorted to battery +. Who is right?

   a. A only

   b. B only

   c. Both A and B

   d. Neither A nor B

3. What step should be taken before beginning a manufacturer's specific testing routine?

   a. Perform basic visual inspections

   b. Check the battery and charging system

   c. Verify MIL is lit or off to choose the diagnostic path.

   d. All of the above

4. An OBD II vehicle has a stored DTC P0300 (Random Misfire). All of the following could cause this *except:*

   a. CKP sensor

   b. Poor ignition primary circuit connection

   c. A defective spark plug

   d. Defective spark plug wires

5. After diagnosing and repairing a malfunction, what should the technician do first?

   a. Perform a road test

   b. Record all available data and erase the DTCs

   c. Observe the scan tool data stream display

   d. Perform any necessary reinitialization procedures

6. If a first trip failure is present for a two-trip code, where will you find it with the scan tool?

    a. Freeze Frame data display

    b. Data stream display

    c. Pending codes display

    d. Readiness Status display

7

Component Testing

We stated earlier that the single most important tool you need is a clear understanding of how the system or circuit you are working with is supposed to work. By understanding the function of components and circuits in the system you are troubleshooting, you can take a system approach to your procedures. In addition, it will be easier to pick the correct diagnostic tool to use and decide exactly what measurements or observations to make.

Purposes of Component Testing

It is essential to understand that all of the diagnostic inspections, troubleshooting and specific routines we have seen so far have one thing in common: they are *responses*, designed to locate the root cause of whatever problem caused the OBD II system to detect and report a failure. Remember that OBD II is not a control system, it performs monitoring and data recording functions only. All control functions are contained within the PCM programs, but are *not* part of OBD II. The tests we will discuss in this section also are designed to either locate a root cause or to confirm information we have already gathered.

107

In general, there are four reasons to test individual components:

1. **To compare input and output measurements to scan tool values**
   Remember that the information displayed by a scan tool has been translated at least twice and updates relatively slowly. Taking measurements directly at a component allows the technician to actually "see" what the computer is receiving or commanding. Thus, we are not forced to rely on what the scan tool thought the PCM said that it thought it saw. This type of measurement is especially useful when checking the actual operation of a suspect sensor, or when evaluating how well the engine management systems are being controlled by the PCM. Signal, reference voltage, ground, and bias voltage values all can be individually measured with the components connected. With a circuit opened, resistance of each component or circuit section may be measured and compared.

2. **To check the quality of connections**   The quality of any component connection or portion of a circuit may be assessed by use of the voltage drop test when the circuit is operating, or by measuring its resistance while it is not operating.

3. **To verify or isolate a problem indicated by diagnostic tests or DTC**
   An experienced auto technician will tell you that a DTC indicating a defective sensor does not necessarily mean that the sensor is the problem. As we have seen by examining examples of DTC diagnostic routines in detail, there can be a wide variety of root causes, all leading to the same DTC being set. In many cases, the quickest way to begin the diagnosis is to look up the normal value or range for the component referred to by the DTC. Then, go directly to that component and take the measurement(s). If the initial test is normal, you still have eliminated the component as the "prime suspect" in causing the problem.

4. **To observe component operation in real time**   Direct testing of a component allows its operation to be watched as it functions. In doing so, its response to changes in engine operating conditions can be tracked, and any "glitches" in its operation can be seen immediately. This is important, because the rate at which a scan tool updates its data and display is relatively slow. Momentary abnormal signals may not be seen if they happen to occur between updates. In addition, a sensor's variable signal may be observed throughout its entire range of operation, so a "dead spot" or "skewed" signal can be spotted to aid in diagnosis.

Before Beginning the Tests

Before you begin component testing, it is essential to have the necessary service information for the vehicle being repaired in order to know what the value you are measuring should be. Never guess at what reading or range is okay and never trust "typical" values. Your vehicle may not be typical. Even a very small variation from normal voltage, pressure, range, or resistance can cause driveability symptoms or cause a DTC to be set.

When performing any tests or repairs that will require disconnecting a power source, removing a fuse, or disconnecting the vehicle battery to avoid the possibility of a short circuit, it is important to think ahead to what systems or components share the same circuit and may be affected. Keep-alive memory may be lost in a variety of components and accessories, some of which may need to be reprogrammed or reset before they will work correctly. Typical systems and components that may require reprogramming include:

- PCM
- Engine throttle servo
- Crankshaft position variation
- Audio system security code
- Memory power seat, pedals, and/or steering wheel tilt system
- Anti-theft system
- Power door opener/closer

Make sure that any required programming or reset codes are known and noted before performing any disconnect procedures. Also, be certain that you have the necessary equipment to perform any needed reprogramming functions. If not, it is best not to disconnect the affected circuit.

In many cases, keep-alive memory can be retained by using a secondary power supply such as a keep-alive device that plugs into the vehicle's cigarette lighter or a portable jump-start unit that can be attached to a portion of the vehicle's circuits with jumper wires. Two types of keep-alive devices are shown in Figure 7–1.

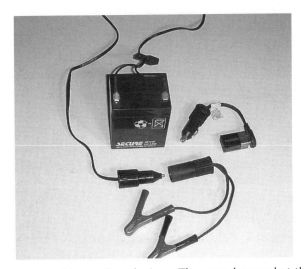

Figure 7–1 Two examples of keep-alive devices. They can be used at the cigarette lighter or other circuit segments.

Note   Some keep-alive devices do not last very long in use. These devices are not intended to be used for more than a few minutes at a time.

Caution   Using such devices may prevent the intended circuit from being disconnected, as well as those you are trying to keep alive. Proceed carefully!

Types of Component Testing

Now, let's look a little closer at some common types of component tests and see how to perform them. We will examine several basic test types.

Voltage Measurement

The digital multimeter is the tool of choice for taking direct voltage measurements. You may want to measure voltage as directed in a diagnostic routine, or to compare the direct measurement to a value shown on the scan tool display. Figure 7–2 shows voltage measurements with a DMM and a scan tool.

Here are some basic rules to follow when testing voltage:

- Voltage measurements are always taken between two specified points of a circuit, *while the circuit is operating*. Measuring available voltage to a component that is not operating can hide excessive resistance in the circuit and give a "false okay" reading.

Figure 7–2   Voltage test for this IAT sensor looks good with the DMM and the scan tool. But, look at Figure 7–7 for the rest of the story!

- If the circuit appears to be open or shorted, it is usually best to start at the component being operated or controlled and work your way back toward the ground connection and the power source in order to isolate the location.

- As a timesaver in troubleshooting a short or open circuit, refer to a component locator guide and wiring diagram. Find the location of the connectors shown in the diagram. If several connectors are used, unplug the connectors one at a time to isolate the open or shorted section of the circuit. Remember to include disconnecting the component to eliminate it as the cause of a short.

- Measuring the reference and/or bias voltage to a suspect sensor may reveal a "hidden" problem in the PCM, especially in the case of intermittent DTCs. First, test the voltage with the ignition on, engine off (KOEO). Then, test it again with the engine running and vary the engine speed. The reference or bias voltage from the PCM is precision regulated to assure accuracy of the sensors and should not fluctuate. If it varies, suspect a PCM problem, but check the charging system first!

- To avoid damaging the insulation of wiring and connectors, do not back-probe or punch holes in wires to take your measurements. Instead, use a nonintrusive, clamp-on type of wire probe such as the one shown in Figure 7–3, and add a jumper wire if necessary. This type of probe penetrates the insulation with a very small pin as the clamp closes, but the hole is so small that it closes over when the clamp is removed.

This type of probe also works with a test light. Simply connect the test light probe to the clamp-on probe with a jumper wire. See Figure 7–4.

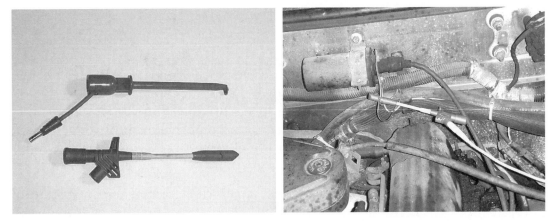

Figure 7–3 Left: Two types of nonintrusive wire harness probes. Right: No harm is done to the insulation with a probe like this one.

Figure 7–4 Even a test light can be used without penetrating the wiring insulation.

Voltage Drop

The voltage drop test is a very efficient way of measuring resistance in a circuit while it is in operation. The reading is taken between two points on a single "leg" of the circuit; either an insulated section or a ground section. While the circuit is in operation, the meter will read any voltage *lost* due to resistance in that circuit section. If no resistance is present, the voltage should read zero, or very close to zero. If available, check the manufacturer's specifications to determine how much voltage drop is permissible in a given circuit. Figure 7–5 is an example of a normal voltage drop reading.

Oscilloscope Testing

For direct component testing, the digital storage oscilloscope (DSO), also known as the *lab scope*, is a very valuable tool. The DSO displays voltage over a period of time. Both the range of voltage and the speed at which time moves across the DSO screen are adjustable. DSOs designed for automotive diagnostics generally have a large menu of adjustments that are preset for testing a wide variety of components and systems, as shown in Figure 7–6.

When connected to the circuit to be tested, a pattern is formed across the screen as the component being tested is operated. The pattern may be captured and stored in the DSO memory, then compared to a known good pattern in a diagnostic software program or service manual. The DSO pattern also may be used in

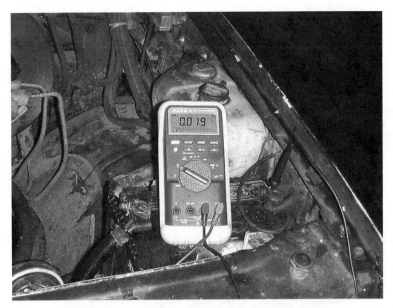

Figure 7–5   As mentioned earlier, a low voltage drop while the circuit is under load means a good connection.

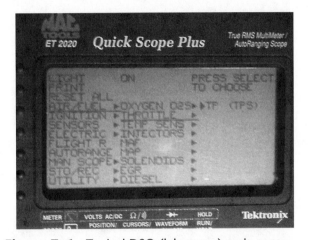

Figure 7–6   Typical DSO (lab scope) main menu.

conjunction with scan tool data to isolate the cause of hard-to-diagnose drive-abiltiy symptoms, such as sensors that may be functioning within range, but are not quite normal.

The DSO is very versatile, capable of being used for advanced levels of engine management system diagnosis. Complete DSO information and instructions are beyond the scope of this course, but the finer points of DSO diagnostic techniques are worth exploring. Many technicians find that use of the DSO saves time and makes accurate diagnosis easier, especially on the "tough jobs." One of the most comprehensive *quick checks* of the entire engine management system is to observe the upstream and downstream HO$_2$S patterns with the DSO. The oxygen sensor pattern gives an overall picture of how efficient the combustion is and how well the PCM is controlling its systems.

Caution   Do not use the DSO and scan tool at the same time to check HO_2S signals. On many vehicles, the PCM goes into a different operating state when it communicates with the scan tool. This may cause false abnormalities in the HO_2S signal pattern. Figures 7–7, 7–8, and 7–9 demonstrate some of the uses of the DSO.

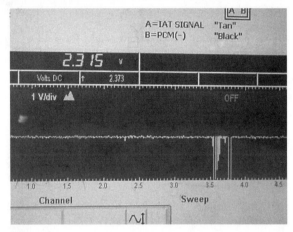

Figure 7–7   The IAT sensor being tested has an intermittent problem! Voltage should be steady, changing slowly in response to temperature. This "hiccup" was missed by the DMM and the scan tool, but it is obvious on the DSO screen.

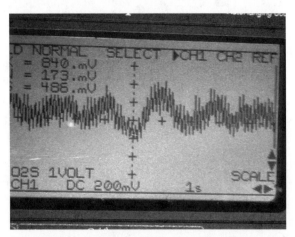

Figure 7–8   Not so good! The HO_2S is switching, but the DSO shows lots of hash on the signal, which would not be picked up by the scan tool. This is a sign of inefficient combustion. Not too surprising, since the engine has clocked 163,000 miles.

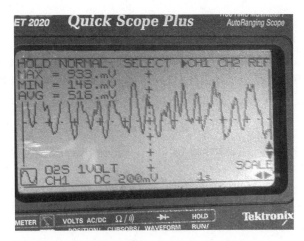

Figure 7–9 Somewhat better at 2500 RPM, but the operating range appears to be "skewed" toward the high end. This condition and the one shown in Figure 7–8 could cause driveabiltiy symptoms or reduced fuel economy and never trigger a DTC or the MIL.

Open Circuit Resistance Testing

For measuring resistance in a component or circuit *when not operating*, the ohmmeter is used. The DMM is still the tool of choice, but it is self-powered when the Ohms function is selected. Damage to the meter could occur if it is connected to a "live" circuit while in the ohmmeter mode.

Resistance may be checked with the ohmmeter between virtually any two points in a circuit or across the connecting terminals of a component, as illustrated in Figure 7–10. It will indicate the amount of resistance between the two test points. Like the voltage drop test, resistance testing can help pinpoint poor connections anywhere in a circuit. If the specifications are known, resistance testing can be helpful in finding short circuits, open circuits, and components that may still be functioning but are not operating normally.

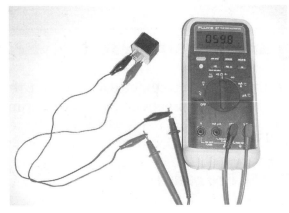

Figure 7–10 Checking the resistance with an ohmmeter will quickly find a defective HO$_2$S heater.

Component Cross-Checking

Cross-checking the operation of inputs and outputs that relate to one another is a good method to use when troubleshooting an intermittent problem or no-code driveability symptom. It is rather like running your own rationality monitor. For examples, let's consider the fuel trim:

- When the EGR is commanded to open, either by using the scan tool or "naturally," the MAP reading should increase (intake manifold vacuum decreases). If the EGR is opened to the point of lean misfire, short-term fuel trim should increase to maintain the correct mixture. If not, the short-term fuel trim may decrease due to less oxygen being present in the exhaust.

- Producing a vacuum leak to the intake system should result in a fuel trim increase, if the leak introduced is large enough to cause lean misfire.

- Plugging the MAP sensor vacuum source and operating it manually with a vacuum pump should produce fuel trim changes and idle quality changes as the vacuum is increased and decreased.

- Loosening the air intake to the throttle body enough to introduce some "false air" downstream of the MAF sensor should cause the fuel trim to decrease as it senses a reduction in the amount of intake air.

- Turning off the fuel pump at warm engine idle should cause the fuel trim to increase as the PCM tries to compensate for the reduced fuel pressure.

Of course, these examples would have other effects as well. Such functions as idle air control, spark advance, exhaust pressure sensor, and oxygen sensor readings would likely be affected. During the process of this type of component testing, the PCM may set DTC and/or turn on the MIL. It is also possible that the system may revert to open loop operation and the effects may not be as expected until closed loop is achieved once again. Be sure to check the details carefully on your scan tool and remember to clear all codes after testing and component tampering are complete.

After Testing and Repair: Resetting Procedures

When testing and repairs have been completed, chances are that one circuit, component, or system will have been disconnected, forced to function, or otherwise disturbed. In some cases, one or more items will have to be reset to function correctly, even after the DTCs are cleared and the repairs have been verified.

Example of a Reset Procedure

The following example is a General Motors Crankshaft Position System Variation Learn Procedure that is essential to allow the PCM to accurately detect misfire in their 2.4 L Quad-4 engine. This learn procedure must be performed any time the crankshaft position sensor-to-crankshaft relationship is changed. Removing the sensor and then reinstalling it is considered to be a disturbance that requires the reset. The procedure also is required with replacement of PCM, engine, CKP sensor, crankshaft, or any engine repair that disturbs the crankshaft, harmonic balancer, or CKP sensor.

GM Crankshaft Position System Variation Learn Procedure

Warning   This procedure requires increasing the engine speed to the programmed rev limit of 3920 RPM. Release the throttle immediately when the engine starts to decelerate in order to prevent over-revving the engine. Once the learn procedure is completed, the control module will return engine control to the operator and the engine will respond to throttle position.

1. Precondition the vehicle, as follows:
   - Set parking brake and block wheels for safety.
   - Ensure that battery is fully charged and in good condition.
   - Verify that scan tool connection to DLC is clean and tight.
2. Close hood.
3. Place transmission gear selector in Park or Neutral.
4. Turn all accessories Off.
5. Connect scan tool.
6. Start and run engine to normal operating temperature of at least 185°F (85°C).
7. With engine still running, enable the CKP learn procedure using scan tool.
8. Depress and hold brake pedal firmly. Raise engine speed to 3920 RPM, releasing the throttle as soon as engine starts to cut out.
9. Using scan tool, verify that crankshaft variation has been learned.

A fully warmed up engine is critical to learning the variation correctly. If a valid learn occurs, no other learn or reset can be completed on the same ignition cycle. If the engine cuts out before the specified engine speed or at normal fuel cutoff RPM, the PCM is not in the learn procedure mode. Review the learn procedure steps and re-enable using the scan tool. Verify that the scan tool displays "Test in Progress." If the variation will not learn, the procedure may be repeated up to 10 times. If scan tool displays an error, the cause is likely to be excessive crankshaft variation. Refer to Figure 7–11 for troubleshooting assistance.

Scan Tool Message	Possible Causes
Factors out of range	Reluctor wheel defect: Machining quality, runout or incorrect air gap
Opposing factors out of range	Electrical disturbance: "Noise" on CKP circuit. Check/reposition wiring, reattempt learn.
Sum out of range	Engine temperature too low. Reattempt learn.
Crank pulse count error	Crank or cam sensor DTC is set. Repair affected DTC first.

Figure 7–11 Excessive crankshaft variation causes.

Power Sliding Door Reset, 2003 Pontiac Montana Minivan

Here is an example of a less critical reset procedure, but one that will be appreciated by your customers. If battery power to the Rear Side Door Actuator Control Module is lost, the power sliding doors will not operate on this model until a reinitialization procedure is performed. Although this loss of function is normal, it may be perceived by the customer as damage caused by the repairs just performed.

Reinitialization Procedure

Author's Note The following procedure refers to checking for DTCs, which may be stored in the control module. These DTCs are not retrieved with the scan tool. Instead, the Power Sliding Door Switch is used to place the system in diagnostic mode. Then, the system alarm buzzes in short bursts to indicate two-digit DTCs. It is similar to counting the MIL flashes on an OBD I vehicle. Unlocking of radio antitheft protection is also mentioned. This is accomplished by entering the radio security code to the Body Control Module (BCM). On some systems, the manufacturer's proprietary scan tool software is required to enter the radio unlocking code.

Manufacturer's Note Before performing Power Sliding Door (PSD) reinitialization, check for DTCs using the self-diagnostic system. Rear side door actuator control modules monitor position of power sliding doors at all times. Whenever battery voltage supply to a rear side door actuator control module has been removed or on-board diagnostics have been activated, power sliding door system(s) must be reinitialized.

> **Caution** If radio is equipped with antitheft protection, radio must be unlocked before performing reinitialization procedure.

1. Turn PSD On/Off switch (on dash at headlight switch assembly) to Off position for system that is to be reinitialized.
2. Manually open power sliding door completely.
3. Manually close and latch power sliding door.
4. Turn power sliding door On/Off switch to On position.
5. Open power sliding door completely using "B" pillar or overhead console Open/Close control switch.
6. Close power sliding door completely using "B" pillar or overhead console Open/Close control switch.

Reset and "Learn" Pitfalls

There are some rather obscure pitfalls with respect to resets and learning procedures that vary from one manufacturer to another. This is another example of situations where the technician may be at a disadvantage if he or she is not equipped with the manufacturer's proprietary diagnostic equipment and software. The following example concerns a joint venture between Ford Motor Company and Nissan, the Mercury Villager and Nissan Quest minivans:

These vehicles provide a separate set of data retrieval and output test functions when the scan tool is connected to the **Diagnostic Data Link (DDL)**, a special connector located inside the fuse box under the dash. The DDL contains a "Work Support" menu function that allows the technician to check and adjust the throttle position sensor and base idle speed. This assures that both TP and IAC are properly adjusted after a repair, and it is important to know that these functions *are not available* using the OBD II standard DLC connection.

Be sure to check the operation of all functions and accessories on the vehicle in the course of your final road test. If something doesn't work correctly, it is likely that a reset procedure is needed. Reset procedures run the gamut. Some involve only the operation of accessories, others are important to allow the PCM and OBD II systems to function properly. Some procedures are quick and easy, others are quite complex. All are worth the effort to make sure that everything on the vehicle works correctly when you return it to the customer. There is no better way to protect your reputation and earn your customer's loyalty.

A Final Note from the Author   Before revision of SAE J 2012 in April, 2002, there were 1000 generic OBD II powertrain trouble codes reserved. With this revision, over 2000 new powertrain DTC code numbers have been reserved. Approximately 1000 are for ISO/SAE (generic) use and 1000 are for a combination of use by manufacturers and reserved for ISO/SAE use. Of those ~2000 new codes, 1500 have already been specifically identified. Many of the new codes identified are related to new, evolving powertrain technologies such as electronically variable valve timing and/or operation, camshaft positioning, reductant injection, and variable intake runner length to name a few. This list sounds overwhelming, but the self-diagnostic capabilities of the PCM also are evolving to become even more sophisticated. The expanded DTC list will make quick, pinpoint diagnosis and repair easier. We will soon see levels of engine efficiency that engineers never even imagined at the beginning of the automotive emission control story. Automotive technicians are poised at the threshold of another new and exciting era in automotive technology. We can literally learn something new every day. I hope that you are as proud and eager as I am to be part of this fascinating transition!

Review and Reinforcement Questions

1. Which of the following tests is best to check an oxygen sensor heater element?

    a. Measure the heater resistance with an ohmmeter.

    b. Test the voltage and ground at the harness connector.

    c. Measure voltage drop on the ground side.

    d. Connect a DSO to the PCM pin 13.

2. With an intermittent sensor DTC, the sensor and circuit check okay with the ohmmeter. Direct component testing shows normal operating range. The PCM is now suspected as the cause. What should the technician do *next*?

    a. Recheck the voltage drop on the ground side.

    b. Measure reference voltage with the engine off and with it running.

    c. Measure the charging system ripple.

    d. Check for power to the sensor with a test light.

3. In the situation described in Question 2 above, what *other* test routine would be best to try to identify the problem?

    a. Direct testing the sensor signal with a DSO while driving the vehicle.

    b. Direct testing the sensor signal with a DMM while driving the vehicle.

   c. Observing the scan tool data for the sensor while driving the vehicle.

   d. Observing the scan tool Parameter ID display for the sensor.

4. The signal voltage of a sensor measures 3.2 volts with a DMM at the connector, but the scan tool shows 2.2 volts. Technician A says a poor connection at the PCM may be causing resistance in the circuit. Technician B says the reference voltage to the sensor may be abnormal. Who is right?

   a. A only

   b. B only

   c. Both A and B

   d. Neither A nor B

5. After replacing the battery in a vehicle, the engine idle is rough. Technician A says to connect the scan tool and look for DTCs. Technician B says to drive the vehicle and allow the PCM to relearn its operating parameters, then reassess the symptom. Who is right?

   a. A only

   b. B only

   c. Both A and B

   d. Neither A nor B

6. Which of the following component signals should *not* be measured with the scan tool and DSO at the same time?

   a. ECT

   b. HO_2S

   c. IAT

   d. TPS

APPENDIX A: Glossary of Automotive Terms and Abbreviations

ACT Air Charge Temperature sensor, monitors temperature of air entering the engine. See also IAT, MAT.

AFM Airflow Meter, measures volume of air entering the engine, may also measure temperature. Less accurate than the Mass Airflow Sensor. See also MAF.

ALDL Assembly Line Data Link, connector for reading trouble codes and engine data, also used to calibrate control systems when the vehicle is assembled. See also DLC.

BARO Barometric Pressure sensor, tells the PCM the atmospheric pressure before the engine is started. The BARO reading is obtained from the Manifold Absolute Pressure sensor on some models. See also MAP.

BCM Body Control Module, computer/processor used to control accessories and functions other than the engine and transmission. Communicates and interacts with the PCM on some models.

BFC Body Functions Computer, see BCM.

BOO Brake On/Off switch, tells the PCM when the brakes are applied and released.

Carbon Monoxide (CO) A tasteless, odorless poisonous gas that is a product of incomplete combustion. In general, CO is produced in the gasoline engine when the air/fuel mixture is too rich. In effect, the combustion stops too soon because there is not enough air to allow it to complete. When a high level of CO is present, no exhaust oxygen should be present. Carbon monoxide is present when the fuel was ignited, but did not finish burning.

Catalytic Converter An emission control device mounted in the exhaust system, which chemically converts harmful pollutants to less harmful substances. The three-way Catalytic Converter treats carbon monoxide, hydrocarbons, and oxides of nitrogen. The catalytic converter has been a main component of exhaust emission controls since 1975.

CCDICM Center Console Driver Information Center Module, information display screen located in the center console that provides direct reading of outputs from the PCM and BCM.

CKP Crankshaft Position sensor, monitors the rotation of the crankshaft and relays the information to the PCM, which uses it to calculate ignition timing and fuel injection pulses. CKP also is used to determine misfire on many models.

Closed Loop The most efficient operating mode of the engine, in which the PCM uses information from all of the signals and actions it is capable of monitoring. Fuel delivery and ignition timing are adjusted to the best possible levels of performance and economy by the PCM, using all available input information.

CMP Camshaft Position sensor, monitors camshaft rotation and relays the information to the PCM, which uses it to supplement the information provided by the CKP sensor. On some models, the CMP signal is needed to decide when to begin ignition and fuel delivery as the engine is first cranked.

CPS See CKP.

CTS Coolant Temperature Sensor, monitors temperature of engine coolant, sends the information to the PCM. CTS input is one measurement involved in calculation of the amount of fuel to inject and spark advance program. See also ECT.

Detonation Knocking or pinging noise noticed during acceleration, caused by hotter than normal combustion temperatures and/or carbon in combustion chambers, which cause unintentional igniting of fuel. When spark plugs also ignite the fuel, the collision of the two "flame fronts" causes a knocking noise and loss of combustion efficiency. May occur with poor EGR flow. Often mistaken for mechanical valve train noise. Severe detonation can cause damage to pistons and valves.

DI Distributor Ignition, generic description for any ignition system that uses a mechanical distributor instead of a CKP and CMP.

DIC Data Information Center, data display located in the dash, center console, or overhead, used to provide readout of data from the PCM and BCM. Same function as the CCDICM and DICM.

DICM Driver Information Center Module, same function as DIC and CCDICM.

DIM Dash Integration Module, used by the Body Control Module (BCM) to control convenience functions such as automatic "memory" preset power seat positions, power mirror positions, and tilt steering wheel positions.

DLC Data Link Connector. Diagnostic Link Connector. Same component as ALDL.

DLL Diagnostic Data Link. A proprietary data communication link, used for specific diagnostic and programming functions. This particular acronym is used by Ford and Nissan. Proprietary functions include adjustment and "learn" procedures for the throttle position sensor and base idle speed.

DMM Digital MultiMeter. A high-impedance test meter used for measuring voltage, amperage, and resistance in electrical circuits and components.

DPFE Differential Pressure Feedback EGR sensor, Ford Motor Company's acronym for EGR valve flow sensor. Measures EGR flow by monitoring the drop in exhaust pressure across a specially designed orifice placed in the system. Reports readings directly to the PCM for use in calculating fuel delivery and ignition timing.

DTC Diagnostic Trouble Code, the number(s) stored in memory by the PCM to indicate a malfunction or abnormal data reading. The presence of stored codes may indicate a fault that will cause the engine to run in less economical, preprogrammed fuel delivery, and spark advance modes.

DVOM Digital Volt-Ohmmeter. See DMM above.

ECT Engine Coolant Temperature sensor, same component as CTS.

EEPROM Electronically Erasable Programmable Read-Only Memory, computer instructions that can be erased and reprogrammed electronically by the manufacturer using the DLC. Used to calibrate the PCM and BCM to match their programs individually to the engine, transmission, and accessories in each vehicle.

EFE Early Fuel Evaporator Valve, partially blocks exhaust flow when the engine is cold to aid in quicker warm-up. Considered an auxiliary emission control, not found on all models. If stuck closed, may cause poor fuel economy and reduced power without setting any trouble code.

EGO Exhaust Gas Oxygen sensor, measures amount of oxygen in the exhaust flow to tell the PCM whether the air/fuel mixture is rich or lean. The PCM will respond by adjusting the amount of fuel delivered by the injectors. The EGO must heat to approximately 600°F before it can work accurately.

EGR Exhaust Gas Recirculation, consists of a valve, operated by the PCM, and an EGR valve position sensor or an exhaust pressure sensor such as a DPFE to measure flow through the valve. Mixes exhaust gas, which has already been burned, into the intake air/fuel mixture. Since the exhaust gas cannot burn again, it functions to reduce temperatures in the combustion chambers. This is one way that high horsepower engines can run on regular, low-octane fuel without detonation.

EI Electronic Ignition, generic description of any distributorless ignition system, where each spark plug is fired directly by the PCM.

Fuel Pressure Regulated by the fuel pump and pressure regulator, specified fuel pressure is critical to assure correct fuel delivery by the injectors. The PCM cannot compensate for incorrect fuel pressure on many models. Incorrect fuel pressure may or may not turn on the MIL or set a DTC.

Fuel Trim Adjustment made by the PCM as needed to add or remove fuel from its basic programmed amounts as the vehicle is driven. Also known as adaptive fuel strategies in earlier models.

Hydrocarbon (HC) A pollutant that occurs when fuel escapes into the exhaust system without being burned. Gasoline that was never ignited in the engine creates a high level of hydrocarbon emissions. This can occur if the cylinder does not seal properly due to a leaking exhaust valve, if the cylinder temperature is too cool and allows the fuel to be quenched without igniting, or if the air/fuel mixture is too lean, so some fuel never ignites.

HEGO Heated Exhaust Gas Oxygen sensor. Function is the same as the EGO sensor, with the addition of a heater to make the sensor begin working quicker after the engine is first started. See also EGO, O_2S, and HO_2S.

HO_2S Heated Oxygen (O_2) Sensor, same component as HEGO. Up to four sensors are used to monitor the amount of oxygen in the engine's exhaust at various test points. Virtually all O_2 sensors used in OBD II vehicles are heated to make them begin working quicker when the engine is first started. See also "Oxygen Sensor."

IAC Idle Air Control, air passages that bypass the throttle are adjusted by the PCM to raise or lower idle speed as needed.

IAT Intake Air Temperature sensor, same as ACT, MAT.

Ignition Timing Adjustable by the PCM, the exact time of firing each spark plug in relation to the position of the crankshaft and valve train. Even slight sensor errors and/or component malfunctions can cause timing to be inaccurate, resulting in loss of power and reduced fuel economy.

Inertia Fuel Shutoff A mechanical switch designed to turn off the fuel pump in the event of an accident or sudden and violent side-to-side motions. These devices have been known to shut off the fuel pump while driving on very rough "washboard" surfaces.

Interface An avenue of communication that involves translating information into an understandable format. With OBD II, the malfunction indicator lamp is the driver interface, while the scan tool is the technician interface.

Knock Sensor Generates an electrical signal to tell the computer when detonation occurs. These sensors are carefully designed to ignore mechanical knocks and other noises. The PCM responds to this signal by retarding the ignition timing on one or more cylinders to eliminate the knocking caused by detonation. See also "Detonation."

LED Light Emitting Diode. A long-lasting electronic device that illuminates when a specified voltage is applied to it. Often used as indicator lights and as components of electronic optical reader devices.

MAF Mass Airflow sensor, used by the PCM to measure the amount and density of intake air. Similar to the airflow meter, but more accurate.

MAP Manifold Absolute Pressure sensor, measures intake manifold vacuum to sense engine load.

MAT Manifold Air Temperature sensor, same as ACT and IAT.

MIL   Malfunction Indicator Lamp, controlled by the PCM, the MIL illuminates to tell the driver that a problem has been detected in one or more of the monitored signals. The MIL displays either a message, such as "check engine," "power loss," "service engine soon," or a small picture of an engine. It may be red or amber, depending on the vehicle make and model. When the MIL is lit, the engine is not likely to be operating at peak efficiency.

Monitor   A procedure performed by the OBD II diagnostic system to actively test the operation of various engine management systems or subsystems that affect exhaust emissions. If a monitor failure occurs, a diagnostic trouble code (DTC) generally will be stored in PCM memory.

MPI or MPFI   Multiport Fuel Injection, currently the most commonly used in which an individual fuel injector is located in or near the intake port of each cylinder.

NO$_x$ See "Oxides of Nitrogen."

Open Loop   Relatively inefficient operating mode of the engine in which the PCM uses preprogrammed calculations to control air/fuel mixture and ignition timing, ignoring a number of actual signals and actions that it observes. The open loop mode exists when the PCM determines that an important signal is out of normal range, or during warm-up, when the engine and oxygen sensors are not yet at normal operating temperature.

Oxides of Nitrogen (NO$_x$)   Oxides of nitrogen are a pollutant that occurs in a gasoline engine when the combustion temperatures are too high. The primary emission control designed to help control NO$_x$ emissions is the Exhaust Gas Recirculation (EGR) system.

Oxygen Sensor (O$_2$S)   An emission control device that generates a direct current (DC) voltage in the *absence* of oxygen, this sensor is used to measure oxygen content in the engine exhaust gases. High oxygen content produces low voltage, near 0 volts, while low oxygen content produces high voltage, usually near 1 volt. If the sensor contains a heater circuit to make it function quicker during warm-up, it is a Heated Oxygen Sensor (HO$_2$S).

PCV   Positive Crankcase Ventilation, the earliest of all auxiliary emission controls, not controlled by the PCM. Uses a mechanical PCV valve and metal and/or rubber lines to capture unburned fuel vapor from the crankcase and direct it into the intake manifold to be burned. A clogged PCV system may not be compensated for by the PCM, especially on older models, and can cause the engine to run too rich, wasting fuel.

PCM   Powertrain Control Module. The computer that controls all engine management and transmission control functions. Also includes on-board self-diagnostic software, such as OBD II.

Scan Tool   Diagnostic electronic reader, used to translate and display engine operating data and stored trouble codes to aid the automotive technician in troubleshooting problems, determining the cause of an illuminated MIL, and testing specific components.

Spark Knock   See "Detonation."

TBI   Throttle Body Injection, uses one or two fuel injectors located above the throttle plates in the throttle body. This design was widely used until the introduction of OBD II. The TBI system is less accurate and therefore less efficient than MPFI.

Three-Way Catalytic Converter   See "Catalytic Converter."

TPS   Throttle Position Sensor, used to tell the PCM exactly how far the driver has opened the throttle by depressing the accelerator.

TCC   Torque Converter Clutch, designed to increase fuel economy by eliminating the 5 to 10 percent slippage normally present in the torque converter of an automatic transmission.

Turbocharger   A horsepower booster that functions by using exhaust gas flow to spin a turbine at high speed during acceleration. The turbine shaft is connected to another wheel that pumps air into the intake manifold under pressure, packing more air/fuel mixture into the cylinders. The result is increased compression and power during acceleration.

VSS   Vehicle Speed Sensor, used to tell the PCM the vehicle's speed. The VSS bases its measurement on transmission output shaft or driveshaft revolutions per minute. On some models, the VSS is used as input for an electronic speedometer instead of a mechanical speedometer cable. The VSS signal also may be used by the antilock brakes, cruise control and traction control systems.

WOT   Wide Open Throttle, a TPS reading, programmed into the PCM's memory, which tells the PCM when the driver has applied full throttle acceleration. When WOT is reached on most models, the PCM turns off the air conditioner compressor and makes adjustments to help the driver accelerate as quickly as possible.

APPENDIX B: SAE Standardized OBD Acronyms

3-2 Timing Solenoid	3-2TS		Battery Positive Voltage	B+
Four-Wheel Drive Low	4X4L		Blower	BLR
Accelerator Pedal	AP		Brake ON/OFF	BOO
Accel. Pedal Position	APP		Brake Pedal Position	BPP
Air Cleaner	ACL		Calculated Load Value	CLV
Air Conditioning	A/C		Camshaft Position	CMP
Air Cond. Clutch	ACC		Canister Purge	CANP
Air Cond. Clutch Switch	ACCS		Carburetor	CARB
Air Cond. Demand	ACD		Central Multipoint Fuel Inj	CMFI
Air Conditioning ON	ACON		Charge Air Cooler	CAC
Ambient Air Temperature	AAT		Closed Loop	CLL
Air Ride Control	ARC		Closed Throttle Position	CTP
Automatic Transaxle	A/T		Clutch Pedal Position	CPP
Automatic Transmission	A/T		Coast Clutch Solenoid	CCS
Barometric Pressure	BARO		Computer-Controlled Dwell	CCD

Constant Control Relay Module	CCRM		Engine Speed	RPM
Continuous Fuel Injection	CFI		Erasable Programmable Read-Only Memory	EPROM
Continuous Trap Oxidizer	CTOX		Evaporative Emission	EVAP
Crankshaft Position	CKP		Exhaust Gas Recirculation	EGR
Critical Flow Venturi	CFV		Exhaust Pressure	EP
Cylinder Identification	CID		Fan Control	FC
Data Link Connector	DLC		Flash Electrically Erasable Programmable Read-Only Mem.	FEEPROM
Data Negative	DATA–			
Data Output Line	DOL		Flexible Fuel	FF
Data Positive	DATA+		Fourth Gear	4GR
Diagnostic Trouble Code	DTC		Freeze Frame	FRZF
Direct Fuel Injection	DFI		Front-Wheel Drive	FWD
Distributor Ignition	DI		Fuel Level Sensor	(none)
Early Fuel Evaporator	EFE		Fuel Pressure	(none)
EGR Temperature	EGRT		Fuel Pump	FP
Electrically Erasable Programmable Read-Only Memory	EEPROM		Fuel Pump Module	FPModule
			Fuel System Status	FuelSYS
Electronic Ignition	EI		Fuel Trim	FT
Engine Control	EC		Generator/Alternator	GEN
Engine Control Module	ECM		Grams per Mile	GPM
Engine Coolant Level	ECL		Ground	GND
Engine Coolant Temperature	ECT		Heated Oxygen Sensor	HO_2S
Engine Modification	EM		Idle Air Control	IAC
Engine Oil Pressure	EOP		Idle Speed Control	ISC
Engine Oil Temperature	EOT		Ignition Control	IC

Ignition Control Module	ICM		Park-Neutral Position	PNP
Indirect Fuel Injection	IFI		Parameter Identification	PID
Inertia Fuel Shutoff	IFS		Positive Crankcase Ventilation	PCV
Input Shaft Speed	ISS		Power Steering Pressure Switch	PSP
Inspection and Maintenance	I/M		Powertrain Control Module	PCM
Intake Air	IA		Programmable Read-Only Memory	PROM
Intake Air Temperature	IAT		Pulsed Secondary Air Injection	PAIR
Intake Manifold Runner Control	IMRC		Pulse Width Modulated	PWM
Knock Sensor	KS		Random Access Memory	RAM
Malfunction Indicator Lamp	MIL		Read-Only Memory	ROM
Manifold Absolute Pressure	MAP		Rear-Wheel Drive	RWD
Manifold Differential Pressure	MDP		Relay Module	RM
Manifold Surface Temperature	MST		Scan Tool	ST
Manifold Vacuum Zone	MVZ		Secondary Air Injection	AIR
Mass Airflow	MAF		Selectable Four-Wheel Drive	S4WD
Mixture Control	MC		Sequential Multipoint Fuel Injection	SFI
Multipoint Fuel Injection	MFI		Service Reminder Indication	SRI
Nonvolatile Random Access Memory	NVRAM		Shift Solenoid	SS
On-Board Diagnostics	OBD		Short Term Fuel Trim	STFT
Open Loop	OL		Smoke Puff Limiter	SPL
Output Shaft Speed	OSS		Spark Advance	SPKADV
Oxidation Catalytic Converter	OC		Supercharger	SC
Oxygen Sensor	O_2S			

Supercharger Bypass	SCB		Trans. Fluid Temperature	TFT
System Readiness Test	SRT		Transmission Range	TR
Third Gear	3GR		Turbine Shaft Speed	TSS
Three-Way Catalyst	TWC		Turbocharger	TC
Throttle Body	TB		Vane Airflow	VAF
Throttle Body Fuel Injection	TBI		Variable Control Relay Module	VCRM
Throttle Actuator Control	TAC		Vehicle Control Module	VCM
Throttle Position	TP		Vehicle Identification Number	VIN
Throttle Position Sensor	TPSensor			
Throttle Position Switch	TPSwitch		Vehicle Speed Sensor	VSS
Torque Converter Clutch	TCC		Voltage Regulator	VR
Torque Converter Clutch Pressure	TCCP		Warm-up Oxidation Catalytic Converter	WU-OC
Transmission Control Module	TCM		Warm-Up Three-Way Catalytic Converter	WU-TWC
Transmission Fluid Pressure	TFP		Wide Open Throttle	WOT

APPENDIX C: Powertrain Diagnostic Trouble Code Definitions, Partial List

Source: SAE Document J-2012

P00XX Fuel and Air Metering and Auxiliary Emission Controls

DTC Number	DTC Naming
P0001	Fuel Volume Regulator Control Circuit/Open
P0002	Fuel Volume Regulator Control Circuit Range/Performance
P0003	Fuel Volume Regulator Control Circuit Low
P0004	Fuel Volume Regulator Control Circuit High
P0005	Fuel Shutoff Valve "A" Control Circuit/Open
P0006	Fuel Shutoff Valve "A" Control Circuit Low
P0007	Fuel Shutoff Valve "A" Control Circuit High
P0008	Engine Position System Performance, Bank 1
P0009	Engine Position System Performance, Bank 2
P0010*	"A" Camshaft Position Actuator Circuit, Bank 1
P0011*	"A" Camshaft Position—Timing Over-Advanced or System Performance, Bank 1
P0012*	"A" Camshaft Position—Timing Over-Retarded, Bank 1
P0013[†]	"B" Camshaft Position—Actuator Circuit, Bank 1
P0014[†]	"B" Camshaft Position—Timing Over-Advanced or System Performance, Bank 1
P0015[†]	"B" Camshaft Position—Timing Over-Retarded, Bank 1
P0016	Crankshaft Position—Camshaft Position Correlation, Bank 1 Sensor A

DTC Number	DTC Naming
P0017	Crankshaft Position—Camshaft Position Correlation, Bank 1 Sensor B
P0018	Crankshaft Position—Camshaft Position Correlation, Bank 2 Sensor A
P0019	Crankshaft Position—Camshaft Position Correlation, Bank 2 Sensor B
P0020*	"A" Camshaft Position Actuator Circuit, Bank 2
P0021*	"A" Camshaft Position—Timing Over-Advanced or System Performance, Bank 2
P0022*	"A" Camshaft Position—Timing Over-Retarded, Bank 2
P0023[†]	"B" Camshaft Position—Actuator Circuit, Bank 2
P0024[†]	"B" Camshaft Position—Timing Over-Advanced or System Performance, Bank 2
P0025[†]	"B" Camshaft Position—Timing Over-Retarded, Bank 2
P0026	Intake Valve Control Solenoid Circuit Range/Performance, Bank 1
P0027	Exhaust Valve Control Solenoid Circuit Range/Performance, Bank 1
P0028	Intake Valve Control Solenoid Circuit Range/Performance, Bank 2
P0029	Exhaust Valve Control Solenoid Circuit Range/Performance, Bank 2
P0030	HO_2S Heater Control Circuit, Bank 1 Sensor 1
P0031	HO_2S Heater Control Circuit Low, Bank 1 Sensor 1
P0032	HO_2S Heater Control Circuit High, Bank 1 Sensor 1
P0033	Turbo Charger Bypass Valve Control Circuit
P0034	Turbo Charger Bypass Valve Control Circuit Low
P0035	Turbo Charger Bypass Valve Control Circuit High
P0036	HO_2S Heater Control Circuit, Bank 1 Sensor 2
P0037	HO_2S Heater Control Circuit Low, Bank 1 Sensor 2
P0038	HO_2S Heater Control Circuit High, Bank 1 Sensor 2
P0039	Turbo/Super Charger Bypass Valve Control Circuit Range/Performance
P0040	O_2 Sensor Signals Swapped, Bank 1 Sensor 1/Bank 2 Sensor 1
P0041	O_2 Sensor Signals Swapped, Bank 1 Sensor 2/Bank 2 Sensor 2
P0042	HO_2S Heater Control Circuit, Bank 1 Sensor 3
P0043	HO_2S Heater Control Circuit Low, Bank 1 Sensor 3
P0044	HO_2S Heater Control Circuit High, Bank 1 Sensor 3
P0045	Turbo/Super Charger Boost Control Solenoid Circuit/Open
P0046	Turbo/Super Charger Boost Control Solenoid Circuit Range/Performance
P0047	Turbo/Super Charger Boost Control Solenoid Circuit Low

DTC Number	DTC Naming
P0048	Turbo/Super Charger Boost Control Solenoid Circuit High
P0049	Turbo/Super Charger Turbine Overspeed
P0050	HO_2S Heater Control Circuit, Bank 2 Sensor 1
P0051	HO_2S Heater Control Circuit Low, Bank 2 Sensor 1
P0052	HO_2S Heater Control Circuit High, Bank 2 Sensor 1
P0053	HO_2S Heater Resistance, Bank 1 Sensor 1
P0054	HO_2S Heater Resistance, Bank 1 Sensor 2
P0055	HO_2S Heater Resistance, Bank 1 Sensor 3
P0056	HO_2S Heater Control Circuit, Bank 2 Sensor 2
P0057	HO_2S Heater Control Circuit Low, Bank 2 Sensor 2
P0058	HO_2S Heater Control Circuit High, Bank 2 Sensor 2
P0059	HO_2S Heater Resistance, Bank 2 Sensor 1
P0060	HO_2S Heater Resistance, Bank 2 Sensor 2
P0061	HO_2S Heater Resistance, Bank 2 Sensor 3
P0062	HO_2S Heater Control Circuit, Bank 2 Sensor 3
P0063	HO_2S Heater Control Circuit Low, Bank 2 Sensor 3
P0064	HO_2S Heater Control Circuit High, Bank 2 Sensor 3
P0065	Air Assisted Injector Control Range/Performance
P0066	Air Assisted Injector Control Circuit or Circuit Low
P0067	Air Assisted Injector Control Circuit High
P0068	MAP/MAF—Throttle Position Correlation
P0069	Manifold Absolute Pressure—Barometric Pressure Correlation
P0070	Ambient Air Temperature Sensor Circuit
P0071	Ambient Air Temperature Sensor Range/Performance
P0072	Ambient Air Temperature Sensor Circuit Low
P0073	Ambient Air Temperature Sensor Circuit High
P0074	Ambient Air Temperature Sensor Circuit Intermittent
P0075	Intake Valve Control Solenoid Circuit, Bank 1
P0076	Intake Valve Control Solenoid Circuit Low, Bank 1
P0077	Intake Valve Control Solenoid Circuit High, Bank 1
P0078	Exhaust Valve Control Solenoid Circuit, Bank 1
P0079	Exhaust Valve Control Solenoid Circuit Low, Bank 1
P0080	Exhaust Valve Control Solenoid Circuit high, Bank 1

DTC Number	DTC Naming
P0081	Intake Valve Control Solenoid Circuit, Bank 2
P0082	Intake Valve Control Solenoid Circuit Low, Bank 2
P0083	Intake Valve Control Solenoid Circuit High, Bank 2
P0084	Exhaust Valve Control Solenoid Circuit, Bank 2
P0085	Exhaust Valve Control Solenoid Circuit Low, Bank 2
P0086	Exhaust Valve Control Solenoid Circuit High, Bank 2
P0087	Fuel Rail/System Pressure—Too Low
P0088	Fuel Rail/System Pressure—Too High
P0089	Fuel Pressure Regulator 1 Performance
P0090	Fuel Pressure Regulator 1 Control Circuit
P0091	Fuel Pressure Regulator 1 Control Circuit Low
P0092	Fuel Pressure Regulator 1 Control Circuit High
P0093	Fuel System Leak Detected—Large Leak
P0094	Fuel System Leak Detected—Small Leak
P0095	Intake Air Temperature Sensor 2 Circuit
P0096	Intake Air Temperature Sensor 2 Circuit Range/Performance
P0097	Intake Air Temperature Sensor 2 Circuit Low
P0098	Intake Air Temperature Sensor 2 Circuit High
P0099	Intake Air Temperature Sensor 2 Circuit Intermittent/Erratic

*The "A" camshaft shall be either the intake, left, or front camshaft. Left/right and front/rear are determined as if viewed from the driver's seating position. Bank 1 contains cylinder number one, Bank 2 is the opposite bank.

†The "B" camshaft shall be either the exhaust, right, or rear camshaft. Left/right and front/rear are determined as if viewed from the driver's seating position. Bank 1 contains cylinder number one, Bank 2 is the opposite bank.

P01XX Fuel and Air Metering

DTC Number	DTC Naming
P0100	Mass or Volume Airflow Circuit
P0101	Mass or Volume Airflow Circuit Range/Performance
P0102	Mass or Volume Airflow Circuit Low Input
P0103	Mass or Volume Airflow Circuit High Input
P0104	Mass or Volume Airflow Circuit Intermittent
P0105	Manifold Absolute Pressure/Barometric Pressure Circuit
P0106	Manifold Absolute Pressure/Barometric Pressure Circuit Range/Performance
P0107	Manifold Absolute Pressure/Barometric Pressure Circuit Low Input
P0108	Manifold Absolute Pressure/Barometric Pressure Circuit High Input
P0109	Manifold Absolute Pressure/Barometric Pressure Circuit Intermittent
P0110	Intake Air Temperature Sensor 1 Circuit
P0111	Intake Air Temperature Sensor 1 Circuit Range/Performance
P0112	Intake Air Temperature Sensor 1 Circuit Low
P0113	Intake Air Temperature Sensor 1 Circuit High
P0114	Intake Air Temperature Sensor 1 Circuit Intermittent
P0115	Engine Coolant Temperature Circuit
P0116	Engine Coolant Temperature Circuit Range/Performance
P0117	Engine Coolant Temperature Circuit Low
P0118	Engine Coolant Temperature Circuit High
P0119	Engine Coolant Temperature Circuit Intermittent
P0120	Throttle/Pedal Position Sensor/Switch "A" Circuit
P0121	Throttle/Pedal Position Sensor/Switch "A" Circuit Range/Performance
P0122	Throttle/Pedal Position Sensor/Switch "A" Circuit Low
P0123	Throttle/Pedal Position Sensor/Switch "A" Circuit High
P0124	Throttle/Pedal Position Sensor/Switch "A" Circuit Intermittent
P0125	Insufficient Coolant Temperature for Closed Loop Fuel Control
P0126	Insufficient Coolant Temperature for Stable Operation
P0127	Intake Air Temperature Too High
P0128	Coolant Thermostat (Coolant Temperature below Thermostat Regulating Temperature)

DTC Number	DTC Naming
P0129	Barometric Pressure Too Low
P0130	O_2 Sensor Circuit, Bank 1 Sensor 1
P0131	O_2 Sensor Circuit Low Voltage, Bank 1 Sensor 1
P0132	O_2 Sensor Circuit High Voltage, Bank 1 Sensor 1
P0133	O_2 Sensor Circuit High Voltage, Bank 1 Sensor 1
P0134	O_2 Sensor Circuit No Activity Detected, Bank 1 Sensor 1
P0135	O_2 Sensor Heater Circuit, Bank 1 Sensor 1
P0136	O_2 Sensor Circuit, Bank 1 Sensor 2
P0137	O_2 Sensor Circuit Low Voltage, Bank 1 Sensor 2
P0138	O_2 Sensor Circuit High Voltage, Bank 1 Sensor 2
P0139	O_2 Sensor Circuit Slow Response, Bank 1 Sensor 2
P0140	O_2 Sensor Circuit No Activity Detected, Bank 1 Sensor 2
P0141	O_2 Sensor Heater Circuit, Bank 1 Sensor 2
P0142	O_2 Sensor Circuit, Bank 1 Sensor 3
P0143	O_2 Sensor Circuit Low Voltage, Bank 1 Sensor 3
P0144	O_2 Sensor Circuit High Voltage, Bank 1 Sensor 3
P0145	O_2 Sensor Circuit Slow Response, Bank 1 Sensor 3
P0146	O_2 Sensor Circuit No Activity Detected, Bank 1 Sensor 3
P0147	O_2 Sensor Heater Circuit, Bank 1, Sensor 3
P0148	Fuel Delivery Error
P0149	Fuel Timing Error
P0150	O_2 Sensor Circuit, Bank 2 Sensor 1
P0151	O_2 Sensor Circuit Low Voltage, Bank 2 Sensor 1
P0152	O_2 Sensor Circuit High Voltage, Bank 2 Sensor 1
P0153	O_2 Sensor Circuit Slow Response, Bank 2 Sensor 1
P0154	O_2 Sensor Circuit No Activity Detected, Bank 2 Sensor 1
P0155	O_2 Sensor Heater Circuit, Bank 2 Sensor 1
P0156	O_2 Sensor Circuit, Bank 2 Sensor 2
P0157	O_2 Sensor Circuit Low Voltage, Bank 2 Sensor 2
P0158	O_2 Sensor Circuit High Voltage, Bank 2 Sensor 2
P0159	O_2 Sensor Circuit Slow Response, Bank 2 Sensor 2
P0160	O_2 Sensor Circuit No Activity Detected, Bank 2 Sensor 2

DTC Number	DTC Naming
P0161	O_2 Sensor Heater Circuit, Bank 2 Sensor 2
P0162	O_2 Sensor Circuit, Bank 2 Sensor 3
P0163	O_2 Sensor Circuit Low Voltage, Bank 2 Sensor 3
P0164	O_2 Sensor Circuit High Voltage, Bank 2 Sensor 3
P0165	O_2 Sensor Circuit Slow Response, Bank 2 Sensor 3
P0166	O_2 Sensor Circuit No Activity Detected, Bank 2 Sensor 3
P0167	O_2 Sensor Heater Circuit, Bank 2 Sensor 3
P0168	Fuel Temperature Too High
P0169	Incorrect Fuel Consumption
P0170	Fuel Trim, Bank 1
P0171	System Too Lean, Bank 1
P0172	System Too Rich, Bank 1
P0173	Fuel Trim, Bank 2
P0174	System Too Lean, Bank 2
P0175	System Too Rich, Bank 2
P0176	Fuel Composition Sensor Circuit
P0177	Fuel Composition Sensor Circuit Range/Performance
P0178	Fuel Composition Sensor Circuit Low
P0179	Fuel Composition Sensor Circuit High
P0180	Fuel Temperature Sensor A Circuit
P0181	Fuel Temperature Sensor A Circuit Range/Performance
P0182	Fuel Temperature Sensor A Circuit Low
P0183	Fuel Temperature Sensor A Circuit High
P0184	Fuel Temperature Sensor A Circuit Intermittent
P0185	Fuel Temperature Sensor B Circuit
P0186	Fuel Temperature Sensor B Circuit Range/Performance
P0187	Fuel Temperature Sensor B Circuit Low
P0188	Fuel Temperature Sensor B Circuit High
P0189	Fuel Temperature Sensor B Circuit Intermittent
P0190	Fuel Rail Pressure Sensor Circuit
P0191	Fuel Rail Pressure Sensor Circuit Range/Performance
P0192	Fuel Rail Pressure Sensor Circuit Low

DTC Number	DTC Naming
P0193	Fuel Rail Pressure Sensor Circuit High
P0194	Fuel Rail Pressure Sensor Circuit Intermittent
P0195	Engine Oil Temperature Sensor
P0196	Engine Oil Temperature Sensor Range/Performance
P0197	Engine Oil Temperature Sensor Low
P0198	Engine Oil Temperature Sensor High
P0199	Engine Oil Temperature Sensor Intermittent

P02XX Fuel and Air Metering

DTC Number	DTC Naming
P0200	Injector Circuit/Open
P0201	Injector Circuit/Open—Cylinder 1
P0202	Injector Circuit/Open—Cylinder 2
P0203	Injector Circuit/Open—Cylinder 3
P0204	Injector Circuit/Open—Cylinder 4
P0205	Injector Circuit/Open—Cylinder 5
P0206	Injector Circuit/Open—Cylinder 6
P0207	Injector Circuit/Open—Cylinder 7
P0208	Injector Circuit/Open—Cylinder 8
P0209	Injector Circuit/Open—Cylinder 9
P0210	Injector Circuit/Open—Cylinder 10
P0211	Injector Circuit/Open—Cylinder 11
P0212	Injector Circuit/Open—Cylinder 12
P0213	Cold Start Injector 1
P0214	Cold Start Injector 2
P0215	Engine Shutoff Solenoid
P0216	Injector/Injection Timing Control Circuit
P0217	Engine Coolant over Temperature Condition
P0218	Transmission Fluid over Temperature Condition
P0219	Engine Overspeed Condition
P0220	Throttle/Pedal Position Sensor/Switch "B" Circuit
P0221	Throttle/Pedal Position Sensor/Switch "B" Circuit Range/Performance
P0222	Throttle/Pedal Position Sensor/Switch "B" Circuit Low
P0223	Throttle/Pedal Position Sensor/Switch "B" Circuit High
P0224	Throttle/Pedal Position Sensor/Switch "B" Circuit Intermittent
P0225	Throttle/Pedal Position Sensor/Switch "C" Circuit
P0226	Throttle/Pedal Position Sensor/Switch "C" Circuit Range/Performance
P0227	Throttle/Pedal Position Sensor/Switch "C" Circuit Low
P0228	Throttle/Pedal Position Sensor/Switch "C" Circuit High
P0229	Throttle/Pedal Position Sensor/Switch "C" Circuit Intermittent
P0230	Fuel Pump Primary Circuit

DTC Number	DTC Naming
P0231	Fuel Pump Primary Circuit Low
P0232	Fuel Pump Primary Circuit High
P0233	Fuel Pump Primary Circuit Intermittent
P0234	Turbo/Super Charger Overboost Condition
P0235	Turbo/Super Charger Boost Sensor "A" Circuit
P0236	Turbo/Super Charger Boost Sensor "A" Circuit Range/Performance
P0237	Turbo/Super Charger Boost Sensor "A" Circuit Low
P0238	Turbo/Super Charger Boost Sensor "A" Circuit High
P0239	Turbo/Super Charger Boost Sensor "B" Circuit
P0240	Turbo/Super Charger Boost Sensor "B" Circuit Range/Performance
P0241	Turbo/Super Charger Boost Sensor "B" Circuit Low
P0242	Turbo/Super Charger Boost Sensor "B" Circuit High
P0243	Turbo/Super Charger Wastegate Solenoid "A"
P0244	Turbo/Super Charger Wastegate Solenoid "A" Range/Performance
P0245	Turbo/Super Charger Wastegate Solenoid "A" Low
P0246	Turbo/Super Charger Wastegate Solenoid "A" High
P0247	Turbo/Super Charger Wastegate Solenoid "B"
P0248	Turbo/Super Charger Wastegate Solenoid "B" Range/Performance
P0249	Turbo/Super Charger Wastegate Solenoid "B" Low
P0250	Turbo/Super Charger Wastegate Solenoid "B" High
P0251	Injection Pump Fuel Metering Control "A" (Cam/Rotor/Injector)
P0252	Injection Pump Fuel Metering Control "A" Range/Performance (Cam/Rotor/Injector)
P0253	Injection Pump Fuel Metering Control "A" Low (Cam/Rotor/Injector)
P0254	Injection Pump Fuel Metering Control "A" High (Cam/Rotor/Injector)
P0255	Injection Pump Fuel Metering Control "A" Intermittent (Cam/Rotor/Injector)
P0256	Injection Pump Fuel Metering Control "B" (Cam/Rotor/Injector)
P0257	Injection Pump Fuel Metering Control "B" Range/Performance (Cam/Rotor/Injector)
P0258	Injection Pump Fuel Metering Control "B" Low (Cam/Rotor/Injector)
P0259	Injection Pump Fuel Metering Control "B" High (Cam/Rotor/Injector)
P0260	Injection Pump Fuel Metering Control "B" Intermittent (Cam/Rotor/Injector)

DTC Number	DTC Naming
P0261	Cylinder 1 Injector Circuit Low
P0262	Cylinder 1 Injector Circuit High
P0263	Cylinder 1 Contribution/Balance
P0264	Cylinder 2 Injector Circuit Low
P0265	Cylinder 2 Injector Circuit High
P0266	Cylinder 2 Contribution/Balance
P0267	Cylinder 3 Injector Circuit Low
P0268	Cylinder 3 Injector Circuit High
P0269	Cylinder 3 Contribution/Balance
P0270	Cylinder 4 Injector Circuit Low
P0271	Cylinder 4 Injector Circuit High
P0272	Cylinder 4 Contribution/Balance
P0273	Cylinder 5 Injector Circuit Low
P0274	Cylinder 5 Injector Circuit High
P0275	Cylinder 5 Contribution/Balance
P0276	Cylinder 6 Injector Circuit Low
P0277	Cylinder 6 Injector Circuit High
P0278	Cylinder 6 Contribution/Balance
P0279	Cylinder 7 Injector Circuit Low
P0280	Cylinder 7 Injector Circuit High
P0281	Cylinder 7 Contribution/Balance
P0282	Cylinder 8 Injector Circuit Low
P0283	Cylinder 8 Injector Circuit High
P0284	Cylinder 8 Contribution/Balance
P0285	Cylinder 9 Injector Circuit Low
P0286	Cylinder 9 Injector Circuit High
P0287	Cylinder 9 Contribution/Balance
P0288	Cylinder 10 Injector Circuit Low
P0289	Cylinder 10 Injector Circuit High
P0290	Cylinder 10 Contribution/Balance
P0291	Cylinder 11 Injector Circuit Low
P0292	Cylinder 11 Injector Circuit High
P0293	Cylinder 11 Contribution/Balance

DTC Number	DTC Naming
P0294	Cylinder 12 Injector Circuit Low
P0295	Cylinder 12 Injector Circuit High
P0296	Cylinder 12 Contribution/Balance
P0297	Vehicle Overspeed Condition
P0298	Engine Oil over Temperature
P0299	Turbo/Super Charger Underboost

P03XX Ignition System or Misfire

DTC Number	DTC Naming
P0300	Random/Multiple Cylinder Misfire Detected
P0301	Cylinder 1 Misfire Detected
P0302	Cylinder 2 Misfire Detected
P0303	Cylinder 3 Misfire Detected
P0304	Cylinder 4 Misfire Detected
P0305	Cylinder 5 Misfire Detected
P0306	Cylinder 6 Misfire Detected
P0307	Cylinder 7 Misfire Detected
P0308	Cylinder 8 Misfire Detected
P0309	Cylinder 9 Misfire Detected
P0310	Cylinder 10 Misfire Detected
P0311	Cylinder 11 Misfire Detected
P0312	Cylinder 12 Misfire Detected
P0313	Misfire Detected with Low Fuel
P0314	Single Cylinder Misfire (Cylinder Not Specified)
P0315	Crankshaft Position System Variation Not Learned
P0316	Engine Misfire Detected on Startup (First 1000 Revolutions)
P0317	Rough Road Hardware Not Present
P0318	Rough Road Sensor "A" Signal Circuit
P0319	Rough Road Sensor "B"
P0320	Ignition/Distributor Engine Speed Input Circuit
P0321	Ignition/Distributor Engine Speed Input Circuit Range/Performance
P0322	Ignition/Distributor Engine Speed Input Circuit No Signal
P0323	Ignition/Distributor Engine Speed Input Circuit Intermittent
P0324	Knock Control System Error
P0325	Knock Sensor 1 Circuit, Bank 1 or Single Sensor
P0326	Knock Sensor 1 Circuit Range/Performance, Bank 1 or Single Sensor
P0327	Knock Sensor 1 Circuit Low, Bank 1 or Single Sensor
P0328	Knock Sensor 1 Circuit High, Bank 1 or Single Sensor
P0329	Knock Sensor 1 Circuit Input Intermittent, Bank 1 or Single Sensor
P0330	Knock Sensor 2 Circuit, Bank 2

DTC Number	DTC Naming
P0331	Knock Sensor 2 Circuit Range/Performance, Bank 2
P0332	Knock Sensor 2 Circuit Low, Bank 2
P0333	Knock Sensor 2 Circuit High, Bank 2
P0334	Knock Sensor 2 Circuit Input Intermittent, Bank 2
P0335	Crankshaft Position Sensor "A" Circuit
P0336	Crankshaft Position Sensor "A" Circuit Range/Performance
P0337	Crankshaft Position Sensor "A" Circuit Low
P0338	Crankshaft Position Sensor "A" Circuit High
P0339	Crankshaft Position Sensor "A" Circuit Intermittent
P0340	Crankshaft Position Sensor "A" Circuit, Bank 1 or Single Sensor
P0341	Crankshaft Position Sensor "A" Circuit Range/Performance, Bank 1 or Single Sensor
P0342	Crankshaft Position Sensor "A" Circuit Low, Bank 1 or Single Sensor
P0343	Crankshaft Position Sensor "A" Circuit High, Bank 1 or Single Sensor
P0344	Crankshaft Position Sensor "A" Circuit Intermittent, Bank 1 or Single Sensor
P0345	Crankshaft Position Sensor "A" Circuit , Bank 2
P0346	Crankshaft Position Sensor "A" Circuit Range/Performance, Bank 2
P0347	Crankshaft Position Sensor "A" Circuit Low, Bank 2
P0348	Crankshaft Position Sensor "A" Circuit High, Bank 2
P0349	Crankshaft Position Sensor "A" Circuit Intermittent, Bank 2
P0350	Ignition Coil Primary/Secondary Circuit
P0351	Ignition Coil "A" Primary/Secondary Circuit
P0352	Ignition Coil "B" Primary/Secondary Circuit
P0353	Ignition Coil "C" Primary/Secondary Circuit
P0354	Ignition Coil "D" Primary/Secondary Circuit
P0355	Ignition Coil "E" Primary/Secondary Circuit
P0356	Ignition Coil "F" Primary/Secondary Circuit
P0357	Ignition Coil "G" Primary/Secondary Circuit
P0358	Ignition Coil "H" Primary/Secondary Circuit
P0359	Ignition Coil "I" Primary/Secondary Circuit
P0360	Ignition Coil "J" Primary/Secondary Circuit
P0361	Ignition Coil "K" Primary/Secondary Circuit

DTC Number	DTC Naming
P0362	Ignition Coil "L" Primary/Secondary Circuit
P0363	Misfire Detected—Fueling Disabled
P0364	Reserved
P0365	Camshaft Position Sensor "B" Circuit, Bank 1
P0366	Camshaft Position Sensor "B" Circuit Range/Performance, Bank 1
P0367	Camshaft Position Sensor "B" Circuit Low, Bank 1
P0368	Camshaft Position Sensor "B" Circuit High, Bank 1
P0369	Camshaft Position Sensor "B" Circuit Intermittent, Bank 1
P0370	Timing Reference High Resolution Signal "A"
P0371	Timing Reference High Resolution Signal "A" Too Many Pulses
P0372	Timing Reference High Resolution Signal "A" Too Few Pulses
P0373	Timing Reference High Resolution Signal "A" Intermittent/Erratic Pulses
P0374	Timing Reference High Resolution Signal "A" No Pulse
P0375	Timing Reference High Resolution Signal "B"
P0376	Timing Reference High Resolution Signal "B" Too Many Pulses
P0377	Timing Reference High Resolution Signal "B" Too Few Pulses
P0378	Timing Reference High Resolution Signal "B" Intermittent/Erratic Pulses
P0379	Timing Reference High Resolution Signal "B" No Pulses
P0380	Glow Plug/Heater Circuit "A"
P0381	Glow Plug/Heater Indicator Circuit
P0382	Glow Plug/Heater Circuit "B"
P0383–P0384	Reserved by document
P0385	Crankshaft Position Sensor "B" Circuit
P0386	Crankshaft Position Sensor "B" Circuit Range/Performance
P0387	Crankshaft Position Sensor "B" Circuit Low
P0388	Crankshaft Position Sensor "B" Circuit High
P0389	Crankshaft Position Sensor "B" Circuit Intermittent
P0390	Crankshaft Position Sensor "B" Circuit, Bank 2
P0391	Crankshaft Position Sensor "B" Circuit Range/Performance, Bank 2
P0392	Crankshaft Position Sensor "B" Circuit Low, Bank 2
P0393	Crankshaft Position Sensor "B" Circuit High, Bank 2
P0394	Crankshaft Position Sensor "B" Circuit Intermittent, Bank 2

P04XX Auxiliary Emission Controls

DTC Number	DTC Naming
P0400	Exhaust Gas Recirculation Flow
P0401	Exhaust Gas Recirculation Flow Insufficient Detected
P0402	Exhaust Gas Recirculation Flow Excessive Detected
P0403	Exhaust Gas Recirculation Control Circuit
P0404	Exhaust Gas Recirculation Control Circuit Range/Performance
P0405	Exhaust Gas Recirculation Sensor "A" Circuit Low
P0406	Exhaust Gas Recirculation Sensor "A" Circuit High
P0407	Exhaust Gas Recirculation Sensor "B" Circuit Low
P0408	Exhaust Gas Recirculation Sensor "B" Circuit High
P0409	Exhaust Gas Recirculation Sensor "A"
P0410	Secondary Air Injection System
P0411	Secondary Air Injection System Incorrect Flow Detected
P0412	Secondary Air Injection System Switching Valve "A" Circuit
P0413	Secondary Air Injection System Switching Valve "A" Circuit Open
P0414	Secondary Air Injection System Switching Valve "A" Circuit Shorted
P0415	Secondary Air Injection System Switching Valve "B" Circuit
P0416	Secondary Air Injection System Switching Valve "B" Circuit Open
P0417	Secondary Air Injection System Switching Valve "B" Circuit Shorted
P0418	Secondary Air Injection System Control "A" Circuit
P0419	Secondary Air Injection System Control "B" Circuit
P0420	Catalyst System Efficiency below Threshold, Bank 1
P0421	Warm-Up Catalyst Efficiency below Threshold, Bank 1
P0422	Main Catalyst Efficiency below Threshold, Bank 1
P0423	Heated Catalyst Efficiency below Threshold, Bank 1
P0424	Heated Catalyst Temperature below Threshold, Bank 1
P0425	Catalyst Temperature Sensor, Bank 1
P0426	Catalyst Temperature Sensor Range/Performance, Bank 1
P0427	Catalyst Temperature Sensor Low, Bank 1
P0428	Catalyst Temperature Sensor High, Bank 1
P0429	Catalyst Heater Control Circuit, Bank 1
P0430	Catalyst System Efficiency below Threshold, Bank 2

DTC Number	DTC Naming
P0431	Warm Up Catalyst Efficiency below Threshold, Bank 2
P0432	Main Catalyst Efficiency below Threshold, Bank 2
P0433	Heated Catalyst Efficiency below Threshold, Bank 2
P0434	Heated Catalyst Temperature below Threshold, Bank 2
P0435	Catalyst Temperature Sensor, Bank 2
P0436	Catalyst Temperature Sensor Range/Performance, Bank 2
P0437	Catalyst Temperature Sensor Low, Bank 2
P0438	Catalyst Temperature Sensor High, Bank 2
P0439	Catalyst Heater Control Circuit, Bank 2
P0440	Evaporative Emission System
P0441	Evaporative Emission System Incorrect Purge Flow
P0442	Evaporative Emission System Leak Detected (Small Leak)
P0443	Evaporative Emission System Purge Control Valve Circuit
P0444	Evaporative Emission System Purge Control Valve Circuit Open
P0445	Evaporative Emission System Purge Control Valve Circuit Shorted
P0446	Evaporative Emission System Vent Control Circuit
P0447	Evaporative Emission System Vent Control Circuit Open
P0448	Evaporative Emission System Vent Control Circuit Shorted
P0449	Evaporative Emission System Vent Valve/Solenoid Circuit
P0450	Evaporative Emission System Pressure Sensor/Switch
P0451	Evaporative Emission System Pressure Sensor/Switch Range/Performance
P0452	Evaporative Emission System Pressure Sensor/Switch Low
P0453	Evaporative Emission System Pressure Sensor/Switch High
P0454	Evaporative Emission System Pressure Sensor/Switch Intermittent
P0455	Evaporative Emission System Leak Detected (Large Leak)
P0456	Evaporative Emission System Leak Detected (Very Small Leak)
P0457	Evaporative Emission System Leak Detected (Fuel Cap Loose/Off)
P0458	Evaporative Emission System Purge Control Valve Circuit Low
P0459	Evaporative Emission System Purge Control Valve Circuit High
P0460	Fuel Level Sensor "A" Circuit
P0461	Fuel Level Sensor "A" Circuit Range/Performance
P0462	Fuel Level Sensor "A" Circuit Low

DTC Number	DTC Naming
P0463	Fuel Level Sensor "A" Circuit High
P0464	Fuel Level Sensor "A" Circuit Intermittent
P0465	EVAP Purge Flow Sensor Circuit
P0466	EVAP Purge Flow Sensor Circuit Range/Performance
P0467	EVAP Purge Flow Sensor Circuit Low
P0468	EVAP Purge Flow Sensor Circuit High
P0469	EVAP Purge Flow Sensor Circuit Intermittent
P0470	Exhaust Pressure Sensor
P0471	Exhaust Pressure Sensor Range/Performance
P0472	Exhaust Pressure Sensor Low
P0473	Exhaust Pressure Sensor High
P0474	Exhaust Pressure Sensor Intermittent
P0475	Exhaust Pressure Control Valve
P0476	Exhaust Pressure Control Valve Range/Performance
P0477	Exhaust Pressure Control Valve Low
P0478	Exhaust Pressure Control Valve High
P0479	Exhaust Pressure Control Valve Intermittent
P0480	Fan 1 Control Circuit
P0481	Fan 2 Control Circuit
P0482	Fan 3 Control Circuit
P0483	Fan Rationality Check
P0484	Fan Circuit over Current
P0485	Fan Power/Ground Circuit
P0486	Exhaust Gas Recirculation Sensor "B" Circuit
P0487	Exhaust Gas Recirculation Throttle Position Control Circuit
P0488	Exhaust Gas Recirculation Throttle Position Control Range/Performance
P0489	Exhaust Gas Recirculation Control Circuit Low
P0490	Exhaust Gas Recirculation Control Circuit High
P0491	Secondary Air Injection System Insufficient Flow, Bank 1
P0492	Secondary Air Injection System Insufficient Flow, Bank 2
P0493	Fan Overspeed
P0494	Fan Speed Low

DTC Number	DTC Naming
P0495	Fan Speed High
P0496	Evaporative Emission System High Purge Flow
P0497	Evaporative Emission System Low Purge Flow
P0498	Evaporative Emission System Vent Valve Control Circuit Low
P0499	Evaporative Emission System Vent Valve Control Circuit High

P05XX Vehicle Speed, Idle Control, and Auxiliary Inputs

DTC Number	DTC Naming
P0500	Vehicle Speed Sensor "A"
P0501	Vehicle Speed Sensor "A" Range/Performance
P0502	Vehicle Speed Sensor "A" Circuit Low Input
P0503	Vehicle Speed Sensor "A" Intermittent/Erratic/High
P0504	Brake Switch "A"/"B" Correlation
P0505	Idle Air Control System
P0506	Idle Air Control System RPM Lower than Expected
P0507	Idle Air Control System RPM Higher than Expected
P0508	Idle Air Control System Circuit Low
P0509	Idle Air Control System Circuit High
P0510	Closed Throttle Position Switch
P0511	Idle Air Control Circuit
P0512	Starter Request Circuit
P0513	Incorrect Immobilizer Key
P0514	Battery Temperature Sensor Circuit Range/Performance
P0515	Battery Temperature Sensor Circuit
P0516	Battery Temperature Sensor Circuit Low
P0517	Battery Temperature Sensor Circuit High
P0518	Idle Air Control Circuit Intermittent
P0519	Idle Air Control System Performance
P0520	Engine Oil Pressure Sensor/Switch Circuit
P0521	Engine Oil Pressure Sensor/Switch Range/Performance
P0522	Engine Oil Pressure Sensor/Switch Low Voltage
P0523	Engine Oil Pressure Sensor/Switch High Voltage
P0524	Engine Oil Pressure Too Low
P0525	Cruise Control Servo Control Circuit Range/Performance
P0526	Fan Speed Sensor Circuit
P0527	Fan Speed Sensor Circuit Range/Performance
P0528	Fan Speed Sensor Circuit No Signal
P0529	Fan Speed Sensor Circuit Intermittent
P0530	A/C Refrigerant Pressure Sensor "A" Circuit

DTC Number	DTC Naming
P0531	A/C Refrigerant Pressure Sensor "A" Circuit Range/Performance
P0532	A/C Refrigerant Pressure Sensor "A" Circuit Low
P0533	A/C Refrigerant Pressure Sensor "A" Circuit High
P0534	Air Conditioner Refrigerant Charge Loss
P0535	A/C Evaporator Temperature Sensor Circuit
P0536	A/C Evaporator Temperature Sensor Circuit Range/Performance
P0537	A/C Evaporator Temperature Sensor Circuit Low
P0538	A/C Evaporator Temperature Sensor Circuit High
P0539	A/C Evaporator Temperature Sensor Circuit Intermittent
P0540*	Intake Air Heater "A" Circuit
P0541*	Intake Air Heater "A" Circuit Low
P0542*	Intake Air Heater "A" Circuit High
P0543*	Intake Air Heater "A" Circuit Open
P0544	Exhaust Gas Temperature Sensor Circuit, Bank 1 Sensor 1
P0545	Exhaust Gas Temperature Sensor Circuit Low, Bank 1 Sensor 1
P0546	Exhaust Gas Temperature Sensor Circuit High, Bank 1 Sensor 1
P0547	Exhaust Gas Temperature Sensor Circuit, Bank 2 Sensor 1
P0548	Exhaust Gas Temperature Sensor Circuit Low, Bank 2 Sensor 1
P0549	Exhaust Gas Temperature Sensor Circuit High, Bank 2 Sensor 1
P0550	Power Steering Pressure Sensor/Switch Circuit
P0551	Power Steering Pressure Sensor/Switch Circuit Range/Performance
P0552	Power Steering Pressure Sensor/Switch Circuit Low Input
P0553	Power Steering Pressure Sensor/Switch Circuit High Input
P0554	Power Steering Pressure Sensor/Switch Circuit Intermittent
P0555	Brake Booster Pressure Sensor Circuit
P0556	Brake Booster Pressure Sensor Circuit Range/Performance
P0557	Brake Booster Pressure Sensor Circuit Low Input
P0558	Brake Booster Pressure Sensor Circuit High Input
P0559	Brake Booster Pressure Sensor Circuit Intermittent
P0560	System Voltage
P0561	System Voltage Unstable
P0562	System Voltage Low
P0563	System Voltage High

DTC Number	DTC Naming
P0564	Cruise Control Multifunction Input "A" Circuit
P0565	Cruise Control On Signal
P0566	Cruise Control Off Signal
P0567	Cruise Control Resume Signal
P0568	Cruise Control Set Signal
P0569	Cruise Control Coast Signal
P0570	Cruise Control Accelerate Signal
P0571	Brake Switch "A" Circuit
P0572	Brake Switch "A" Circuit Low
P0573	Brake Switch "A" Circuit High
P0574	Cruise Control System—Vehicle Speed Too High
P0575	Cruise Control Input Circuit
P0576	Cruise Control Input Circuit Low
P0577	Cruise Control Input Circuit High
P0578[†]	Cruise Control Multifunction Input "A" Circuit Stuck
P0579[†]	Cruise Control Multifunction Input "A" Circuit Range/Performance
P0580[†]	Cruise Control Multifunction Input "A" Circuit Low
P0581[†]	Cruise Control Multifunction Input "A" Circuit High
P0582	Cruise Control Vacuum Control Circuit/Open
P0583	Cruise Control Vacuum Control Circuit Low
P0584	Cruise Control Vacuum Control Circuit High
P0585	Cruise Control Multifunction Input "A"/"B" Correlation
P0586	Cruise Control Vent Control Circuit/Open
P0587	Cruise Control Vent Control Circuit Low
P0588	Cruise Control Vent Control Circuit High
P0589	Cruise Control Multifunction Input "B" Circuit
P0590	Cruise Control Multifunction Input "B" Circuit Stuck
P0591	Cruise Control Multifunction Input "B" Circuit Range/Performance
P0592	Cruise Control Multifunction Input "B" Circuit Low
P0593	Cruise Control Multifunction Input "B" Circuit High
P0594	Cruise Control Servo Control Circuit/Open
P0595	Cruise Control Servo Control Circuit Low

DTC Number	DTC Naming
P0596	Cruise Control Servo Control Circuit High
P0597	Thermostat Heater Control Circuit/Open
P0598	Thermostat Heater Control Circuit Low
P0599	Thermostat Heater Control Circuit High

*For DTCs P0540–P0543 also see P2604–P2609
†For DTCs P0578–P0581 also see P0564

P06XX Computer and Auxiliary Outputs

DTC Number	DTC Naming
P0600	Serial Communication Link
P0601	Internal Control Module Memory Check Sum Error
P0602	Control Module Programming Error
P0603	Internal Control Module Keep Alive Memory (KAM) Error
P0604	Internal Control Module Random Access Memory (RAM) Error
P0605	Internal Control Module Read Only Memory (ROM) Error
P0606	ECM/PCM Processor
P0607	Control Module Performance
P0608	Control Module VSS Output "A"
P0609	Control Module VSS Output "B"
P0610	Control Module Vehicle Options Error
P0611	Fuel Injector Control Module Performance
P0612	Fuel Injector Control Module Relay Control
P0613	TCM Processor
P0614	ECM/TCM Incompatible
P0615	Starter Relay Circuit
P0616	Starter Relay Circuit Low
P0617	Starter Relay Circuit High
P0618	Alternative Fuel Control Module KAM Error
P0619	Alternative Fuel Control RAM/ROM Error
P0620	Generator Control Circuit
P0621	Generator Lamp/L Terminal Circuit
P0622	Generator Field/F Terminal Circuit
P0623	Generator Lamp Control Circuit
P0624	Fuel Cap Lamp Control Circuit
P0625	Generator Field/F Terminal Circuit Low
P0626	Generator Field/F Terminal Circuit High
P0627	Fuel Pump "A" Control Circuit/Open
P0628	Fuel Pump "A" Control Circuit Low
P0629	Fuel Pump "A" Control Circuit High
P0630	VIN Not Programmed or Incompatible—ECM/PCM

DTC Number	DTC Naming
P0631	VIN Not Programmed or Incompatible—TCM
P0632	Odometer Not Programmed—ECM/PCM
P0633	Immobilizer Key Not Programmed—ECM/PCM
P0634	PCM/ECM/TCM Internal Temperature Too High
P0635	Power Steering Control Circuit
P0636	Power Steering Control Circuit Low
P0637	Power Steering Control Circuit High
P0638	Throttle Actuator Control Range/Performance, Bank 1
P0639	Throttle Actuator Control Range/Performance, Bank 2
P0640	Intake Air Heater Control Circuit
P0641	Sensor Reference Voltage "A" Circuit/Open
P0642	Sensor Reference Voltage "A" Low
P0643	Sensor Reference Voltage "A" High
P0644	Driver Display Serial Communication Circuit
P0645	A/C Clutch Relay Control Circuit
P0646	A/C Clutch Relay Control Circuit Low
P0647	A/C Clutch Relay Control Circuit High
P0648	Immobilizer Lamp Control Circuit
P0649	Speed Control Lamp Control Circuit
P0650	Malfunction Indicator Lamp (MIL) Control Circuit
P0651	Sensor Reference Voltage "B" Circuit/Open
P0652	Sensor Reference Voltage "B" Circuit Low
P0653	Sensor Reference Voltage "B" Circuit High
P0654	Engine RPM Output Circuit
P0655	Engine Hot Lamp Output Control Circuit
P0656	Fuel Level Output Circuit
P0657	Actuator Supply Voltage "A" Circuit/Open
P0658	Actuator Supply Voltage "A" Circuit Low
P0659	Actuator Supply Voltage "A" Circuit High
P0660	Intake Manifold Tuning Valve Control Circuit/Open, Bank 1*
P0661	Intake Manifold Tuning Valve Control Circuit Low, Bank 1*
P0662	Intake Manifold Tuning Valve Control Circuit High, Bank 1*
P0663	Intake Manifold Tuning Valve Control Circuit/Open, Bank 2*

DTC Number	DTC Naming
P0664	Intake Manifold Tuning Valve Control Circuit Low, Bank 2*
P0665	Intake Manifold Tuning Valve Control Circuit High, Bank 2*
P0666	PCM/ECM/TCM Internal Temperature Sensor Circuit
P0667	PCM/ECM/TCM Internal Temperature Sensor Range/Performance
P0668	PCM/ECM/TCM Internal Temperature Sensor Circuit Low
P0669	PCM/ECM/TCM Internal Temperature Sensor Circuit High
P0670	Glow Plug Module Control Circuit
P0671	Cylinder 1 Glow Plug Circuit
P0672	Cylinder 2 Glow Plug Circuit
P0673	Cylinder 3 Glow Plug Circuit
P0674	Cylinder 4 Glow Plug Circuit
P0675	Cylinder 5 Glow Plug Circuit
P0676	Cylinder 6 Glow Plug Circuit
P0677	Cylinder 7 Glow Plug Circuit
P0678	Cylinder 8 Glow Plug Circuit
P0679	Cylinder 9 Glow Plug Circuit
P0680	Cylinder 10 Glow Plug Circuit
P0681	Cylinder 11 Glow Plug Circuit
P0682	Cylinder 12 Glow Plug Circuit
P0683	Glow Plug Control Module to PCM Communication Circuit
P0684	Glow Plug Control Module to PCM Communication Circuit Range/Performance
P0685	ECM/PCM Power Relay Control Circuit/Open
P0686	ECM/PCM Power Relay Control Circuit Low
P0687	ECM/PCM Power Relay Control Circuit High
P0688	ECM/PCM Power Relay Sense Circuit/Open
P0689	ECM/PCM Power Relay Sense Circuit Low
P0690	ECM/PCM Power Relay Sense Circuit High
P0691	Fan 1 Control Circuit Low
P0692	Fan 1 Control Circuit High
P0693	Fan 2 Control Circuit Low
P0694	Fan 2 Control Circuit High
P0695	Fan 3 Control Circuit Low

DTC Number	DTC Naming
P0696	Fan 3 Control Circuit High
P0697	Sensor Reference Voltage "C" Circuit/Open
P0698	Sensor Reference Voltage "C" Circuit Low
P0699	Sensor Reference Voltage "C" Circuit High

*DTC Application note for intake manifold tuning valves and intake manifold runner controls:

Active controls are used to modify or control airflow within the engine air intake system. These controls may be used to enhance or modify in-cylinder airflow motion (charge motion), modify the airflow dynamics (manifold tuning) within the intake manifold or both.

Devices that control charge motion are commonly called intake manifold runner control, swirl control valve, and charge motion control valve. The SAE recommended term for any device that controls charge motion is intake manifold runner control (IMRC).

Devices that control manifold dynamics or manifold tuning are commonly called intake manifold tuning valve, long/short runner control, and intake manifold communication control. The SAE recommended term for any device that controls manifold tuning is intake manifold tuning (MT) valve.

P07XX Transmission

DTC Number	DTC Naming
P0700	Transmission Control System (MIL Request)
P0701	Transmission Control System Range/Performance
P0702	Transmission Control System Electrical
P0703	Brake Switch "B" Circuit
P0704	Clutch Switch Input Circuit Malfunction
P0705	Transmission Range Sensor Circuit Malfunction (PRNDL Input)
P0706	Transmission Range Sensor Circuit Range/Performance
P0707	Transmission Range Sensor Circuit Low
P0708	Transmission Range Sensor Circuit High
P0709	Transmission Range Sensor Circuit Intermittent
P0710	Transmission Fluid Temperature Sensor "A" Circuit
P0711	Transmission Fluid Temperature Sensor "A" Circuit Range/Performance
P0712	Transmission Fluid Temperature Sensor "A" Circuit Low
P0713	Transmission Fluid Temperature Sensor "A" Circuit High
P0714	Transmission Fluid Temperature Sensor "A" Circuit Intermittent
P0715	Input/Turbine Speed Sensor "A" Circuit
P0716	Input/Turbine Speed Sensor "A" Circuit Range/Performance
P0717	Input/Turbine Speed Sensor "A" Circuit No Signal
P0718	Input/Turbine Speed Sensor "A" Circuit Intermittent
P0719	Brake Switch "B" Circuit Low
P0720	Output Speed Sensor Circuit
P0721	Output Speed Sensor Circuit Range/Performance
P0722	Output Speed Sensor Circuit No Signal
P0723	Output Speed Sensor Circuit Intermittent
P0724	Brake Switch "B" Circuit High
P0725	Engine Speed Input Circuit
P0726	Engine Speed Input Circuit Range/Performance
P0727	Engine Speed Input Circuit No Signal
P0728	Engine Speed Input Circuit Intermittent
P0729	Gear 6 Incorrect Ratio
P0730	Incorrect Gear Ratio

DTC Number	DTC Naming
P0731	Gear 1 Incorrect Ratio
P0732	Gear 2 Incorrect Ratio
P0733	Gear 3 Incorrect Ratio
P0734	Gear 4 Incorrect Ratio
P0735	Gear 5 Incorrect Ratio
P0736	Reverse Incorrect Ratio
P0737	TCM Engine Speed Output Circuit
P0738	TCM Engine Speed Output Circuit Low
P0739	TCM Engine Speed Output Circuit High
P0740	Torque Converter Clutch Circuit/Open
P0741	Torque Converter Clutch Circuit Performance or Stuck Off
P0742	Torque Converter Clutch Circuit Stuck On
P0743	Torque Converter Clutch Circuit Electrical
P0744	Torque Converter Clutch Circuit Intermittent
P0745	Pressure Control Solenoid "A"
P0746	Pressure Control Solenoid "A" Performance or Stuck Off
P0747	Pressure Control Solenoid "A" Stuck On
P0748	Pressure Control Solenoid "A" Electrical
P0749	Pressure Control Solenoid "A" Intermittent
P0750	Shift Solenoid "A"
P0751	Shift Solenoid "A" Performance or Stuck Off
P0752	Shift Solenoid "A" Stuck On
P0753	Shift Solenoid "A" Electrical
P0754	Shift Solenoid "A" Intermittent
P0755	Shift Solenoid "B"
P0756	Shift Solenoid "B" Performance or Stuck Off
P0757	Shift Solenoid "B" Stuck On
P0758	Shift Solenoid "B" Electrical
P0759	Shift Solenoid "B" Intermittent
P0760	Shift Solenoid "C"
P0761	Shift Solenoid "C" Performance or Stuck Off
P0762	Shift Solenoid "C" Stuck On
P0763	Shift Solenoid "C" Electrical

DTC Number	DTC Naming
P0764	Shift Solenoid "C" Intermittent
P0765	Shift Solenoid "D"
P0766	Shift Solenoid "D" Performance or Stuck Off
P0767	Shift Solenoid "D" Stuck On
P0768	Shift Solenoid "D" Electrical
P0769	Shift Solenoid "D" Intermittent
P0770	Shift Solenoid "E"
P0771	Shift Solenoid "E" Performance or Stuck Off
P0772	Shift Solenoid "E" Stuck On
P0773	Shift Solenoid "E" Electrical
P0774	Shift Solenoid "E" Intermittent
P0775	Pressure Control Solenoid "B"
P0776	Pressure Control Solenoid "B" Performance or Stuck Off
P0777	Pressure Control Solenoid "B" Stuck On
P0778	Pressure Control Solenoid "B" Electrical
P0779	Pressure Control Solenoid "B" Intermittent
P0780	Shift Error
P0781	1-2 Shift
P0782	2-3 Shift
P0783	3-4 Shift
P0784	4-5 Shift
P0785	Shift/Timing Solenoid
P0786	Shift/Timing Solenoid Range/Performance
P0787	Shift/Timing Solenoid Low
P0788	Shift/Timing Solenoid High
P0789	Shift/Timing Solenoid Intermittent
P0790	Normal/Performance Switch Circuit
P0791	Intermediate Shaft Speed Sensor "A" Circuit
P0792	Intermediate Shaft Speed Sensor "A" Circuit Range/Performance
P0793	Intermediate Shaft Speed Sensor "A" Circuit No Signal
P0794	Intermediate Shaft Speed Sensor "A" Circuit Intermittent
P0795	Pressure Control Solenoid "C"

DTC Number	DTC Naming
P0796	Pressure Control Solenoid "C" Performance or Stuck Off
P0797	Pressure Control Solenoid "C" Stuck On
P0798	Pressure Control Solenoid "C" Electrical
P0799	Pressure Control Solenoid "C" Intermittent

P08XX Transmission

DTC Number	DTC Naming
P0800	Transfer Case Control System (MIL Request)
P0801	Reverse Inhibit Control Circuit
P0802	Transmission Control System MIL Request Circuit/Open
P0803	1-4 Upshift (Skip Shift) Solenoid Control Circuit
P0804	1-4 Upshift (Skip Shift) Lamp Control Circuit
P0805	Clutch Position Sensor Circuit
P0806	Clutch Position Sensor Circuit Range/Performance
P0807	Clutch Position Sensor Circuit Low
P0808	Clutch Position Sensor Circuit High
P0809	Clutch Position Sensor Circuit Intermittent
P0810	Clutch Position Control Error
P0811	Excessive Clutch Slippage
P0812	Reverse Input Circuit
P0813	Reverse Output Circuit
P0814	Transmission Range Display Circuit
P0815	Upshift Switch Circuit
P0816	Downshift Switch Circuit
P0817	Starter Disable Circuit
P0818	Driveline Disconnect Switch Input Circuit
P0819	Up and Down Shift Switch to Transmission Range Correlation
P0820	Gear Lever X-Y Position Sensor Circuit
P0821	Gear Lever X Position Circuit
P0822	Gear Lever Y Position Circuit
P0823	Gear Lever X Position Circuit Intermittent
P0824	Gear Lever Y Position Circuit Intermittent
P0825	Gear Lever Push-Pull Switch (Shift Anticipate)
P0826	Up and Down Shift Switch Circuit
P0827	Up and Down Shift Switch Circuit Low
P0828	Up and Down Shift Switch Circuit High
P0829	5-6 Shift
P0830	Clutch Pedal Shift "A" Circuit

DTC Number	DTC Naming
P0831	Clutch Pedal Shift "A" Circuit Low
P0832	Clutch Pedal Shift "A" Circuit High
P0833	Clutch Pedal Shift "B" Circuit
P0834	Clutch Pedal Shift "B" Circuit Low
P0835	Clutch Pedal Shift "B" Circuit High
P0836	Four-Wheel Drive (4WD) Switch Circuit
P0837	Four-Wheel Drive (4WD) Switch Circuit Range/Performance
P0838	Four-Wheel Drive (4WD) Switch Circuit Low
P0839	Four-Wheel Drive (4WD) Switch Circuit High
P0840	Transmission Fluid Pressure Sensor/Switch "A" Circuit
P0841	Transmission Fluid Pressure Sensor/Switch "A" Circuit Range/Performance
P0842	Transmission Fluid Pressure Sensor/Switch "A" Circuit Low
P0843	Transmission Fluid Pressure Sensor/Switch "A" Circuit High
P0844	Transmission Fluid Pressure Sensor/Switch "A" Circuit Intermittent
P0845	Transmission Fluid Pressure Sensor/Switch "B" Circuit
P0846	Transmission Fluid Pressure Sensor/Switch "B" Circuit Range/Performance
P0847	Transmission Fluid Pressure Sensor/Switch "B" Circuit Low
P0848	Transmission Fluid Pressure Sensor/Switch "B" Circuit High
P0849	Transmission Fluid Pressure Sensor/Switch "B" Circuit Intermittent
P0850	Park/Neutral Switch Input Circuit
P0851	Park/Neutral Switch Input Circuit Low
P0852	Park/Neutral Switch Input Circuit High
P0853	Drive Switch Input Circuit
P0854	Drive Switch Input Circuit Low
P0855	Drive Switch Input Circuit High
P0856	Traction Control Input Signal
P0857	Traction Control Input Signal Range/Performance
P0858	Traction Control Input Signal Low
P0859	Traction Control Input Signal High
P0860	Gear Shift Module Communication Circuit
P0861	Gear Shift Module Communication Circuit Low

DTC Number	DTC Naming
P0862	Gear Shift Module Communication Circuit High
P0863	TCM Communication Circuit
P0864	TCM Communication Circuit Range/Performance
P0865	TCM Communication Circuit Low
P0866	TCM Communication Circuit High
P0867	Transmission Fluid Pressure
P0868	Transmission Fluid Pressure Low
P0869	Transmission Fluid Pressure High
P0870	Transmission Fluid Pressure Sensor/Switch "C" Circuit
P0871	Transmission Fluid Pressure Sensor/Switch "C" Circuit Range/Performance
P0872	Transmission Fluid Pressure Sensor/Switch "C" Circuit Low
P0873	Transmission Fluid Pressure Sensor/Switch "C" Circuit High
P0874	Transmission Fluid Pressure Sensor/Switch "C" Circuit Intermittent
P0875	Transmission Fluid Pressure Sensor/Switch "D" Circuit
P0876	Transmission Fluid Pressure Sensor/Switch "D" Circuit Range/Performance
P0877	Transmission Fluid Pressure Sensor/Switch "D" Circuit Low
P0878	Transmission Fluid Pressure Sensor/Switch "D" Circuit High
P0879	Transmission Fluid Pressure Sensor/Switch "D" Circuit Intermittent
P0880	TCM Power Input Signal
P0881	TCM Power Input Signal Range/Performance
P0882	TCM Power Input Signal Low
P0883	TCM Power Input Signal High
P0884	TCM Power Input Signal Intermittent
P0885	TCM Power Relay Control Circuit/Open
P0886	TCM Power Relay Control Circuit Low
P0887	TCM Power Relay Control Circuit High
P0888	TCM Power Relay Sense Circuit
P0889	TCM Power Relay Sense Circuit Range/Performance
P0890	TCM Power Relay Sense Circuit Low
P0891	TCM Power Relay Sense Circuit High
P0892	TCM Power Relay Sense Circuit Intermittent

DTC Number	DTC Naming
P0893	Multiple Gears Engaged
P0894	Transmission Component Slipping
P0895	Shift Time Too Short
P0896	Shift Time Too Long
P0897	Transmission Fluid Deteriorated
P0898	Transmission Control System MIL Request Circuit Low
P0899	Transmission Control System MIL Request Circuit High

P09XX Transmission

DTC Number	DTC Naming
P0900	Clutch Actuator Circuit/Open
P0901	Clutch Actuator Circuit Range/Performance
P0902	Clutch Actuator Circuit Low
P0903	Clutch Actuator Circuit High
P0904	Gate Select Position Circuit
P0905	Gate Select Position Circuit Range/Performance
P0906	Gate Select Position Circuit Low
P0907	Gate Select Position Circuit High
P0908	Gate Select Position Circuit Intermittent
P0909	Gate Select Control Error
P0910	Gate Select Actuator Circuit/Open
P0911	Gate Select Actuator Circuit Range/Performance
P0912	Gate Select Actuator Circuit Low
P0913	Gate Select Actuator Circuit High
P0914	Gear Shift Position Circuit
P0915	Gear Shift Position Circuit Range/Performance
P0916	Gear Shift Position Circuit Low
P0917	Gear Shift Position Circuit High
P0918	Gear Shift Position Circuit Intermittent
P0919	Gear Shift Position Control Error
P0920	Gear Shift Forward Actuator Circuit/Open
P0921	Gear Shift Forward Actuator Circuit Range/Performance
P0922	Gear Shift Forward Actuator Circuit Low
P0923	Gear Shift Forward Actuator Circuit High
P0924	Gear Shift Reverse Actuator Circuit/Open
P0925	Gear Shift Reverse Actuator Circuit Range/Performance
P0926	Gear Shift Reverse Actuator Circuit Low
P0927	Gear Shift Reverse Actuator Circuit High
P0928	Gear Shift Lock Solenoid Control Circuit/Open
P0929	Gear Shift Lock Solenoid Control Circuit Range/Performance
P0930	Gear Shift Lock Solenoid Control Circuit Low

DTC Number	DTC Naming
P0931	Gear Shift Lock Solenoid Control Circuit High
P0932	Hydraulic Pressure Sensor Circuit
P0933	Hydraulic Pressure Sensor Circuit Range/Performance
P0934	Hydraulic Pressure Sensor Circuit Low
P0935	Hydraulic Pressure Sensor Circuit High
P0936	Hydraulic Pressure Sensor Circuit Intermittent
P0937	Hydraulic Oil Temperature Sensor Circuit
P0938	Hydraulic Oil Temperature Sensor Circuit Range/Performance
P0939	Hydraulic Oil Temperature Sensor Circuit Low
P0940	Hydraulic Oil Temperature Sensor Circuit High
P0941	Hydraulic Oil Temperature Sensor Circuit Intermittent
P0942	Hydraulic Pressure Unit
P0943	Hydraulic Pressure Unit Cycling Period Too Short
P0944	Hydraulic Pressure Unit Loss of Pressure
P0945	Hydraulic Pump Relay Circuit/Open
P0946	Hydraulic Pump Relay Circuit Range/Performance
P0947	Hydraulic Pump Relay Circuit Low
P0948	Hydraulic Pump Relay Circuit High
P0949	Auto Shift Manual Adaptive Learning Not Complete
P0950	Auto Shift Manual Control Circuit
P0951	Auto Shift Manual Control Circuit Range/Performance
P0952	Auto Shift Manual Control Circuit Low
P0953	Auto Shift Manual Control Circuit High
P0954	Auto Shift Manual Control Circuit Intermittent
P0955	Auto Shift Manual Mode Circuit
P0956	Auto Shift Manual Mode Circuit Range/Performance
P0957	Auto Shift Manual Mode Circuit Low
P0958	Auto Shift Manual Mode Circuit High
P0959	Auto Shift Manual Mode Circuit Intermittent
P0960	Pressure Control Solenoid "A" Control Circuit/Open
P0961	Pressure Control Solenoid "A" Control Circuit Range/Performance
P0962	Pressure Control Solenoid "A" Control Circuit Low
P0963	Pressure Control Solenoid "A" Control Circuit High

DTC Number	DTC Naming
P0964	Pressure Control Solenoid "B" Control Circuit/Open
P0965	Pressure Control Solenoid "B" Control Circuit Range/Performance
P0966	Pressure Control Solenoid "B" Control Circuit Low
P0967	Pressure Control Solenoid "B" Control Circuit High
P0968	Pressure Control Solenoid "C" Control Circuit/Open
P0969	Pressure Control Solenoid "C" Control Circuit Range/Performance
P0970	Pressure Control Solenoid "C" Control Circuit Low
P0971	Pressure Control Solenoid "C" Control Circuit High
P0972	Shift Solenoid "A" Control Circuit Range/Performance
P0973	Shift Solenoid "A" Control Circuit Low
P0974	Shift Solenoid "A" Control Circuit High
P0975	Shift Solenoid "B" Control Circuit Range/Performance
P0976	Shift Solenoid "B" Control Circuit Low
P0977	Shift Solenoid "B" Control Circuit High
P0978	Shift Solenoid "C" Control Circuit Range/Performance
P0979	Shift Solenoid "C" Control Circuit Low
P0980	Shift Solenoid "C" Control Circuit High
P0981	Shift Solenoid "D" Control Circuit Range/Performance
P0982	Shift Solenoid "D" Control Circuit Low
P0983	Shift Solenoid "D" Control Circuit High
P0984	Shift Solenoid "E" Control Circuit Range/Performance
P0985	Shift Solenoid "E" Control Circuit Low
P0986	Shift Solenoid "E" Control Circuit High
P0987	Transmission Fluid Pressure Sensor/Switch "E" Circuit
P0988	Transmission Fluid Pressure Sensor/Switch "E" Circuit Range/Performance
P0989	Transmission Fluid Pressure Sensor/Switch "E" Circuit Low
P0990	Transmission Fluid Pressure Sensor/Switch "E" Circuit High
P0991	Transmission Fluid Pressure Sensor/Switch "E" Circuit Intermittent
P0992	Transmission Fluid Pressure Sensor/Switch "F" Circuit
P0993	Transmission Fluid Pressure Sensor/Switch "F" Circuit Range/Performance
P0994	Transmission Fluid Pressure Sensor/Switch "F" Circuit Low

DTC Number	DTC Naming
P0995	Transmission Fluid Pressure Sensor/Switch "F" Circuit High
P0996	Transmission Fluid Pressure Sensor/Switch "F" Circuit Intermittent
P0997	Shift Solenoid "F" Control Circuit Range/Performance
P0998	Shift Solenoid "F" Control Circuit Low
P0999	Shift Solenoid "F" Control Circuit High

P0AXX Hybrid Propulsion

DTC Number	DTC Naming
P0A00	Motor Electronics Coolant Temperature Sensor Circuit
P0A01	Motor Electronics Coolant Temperature Sensor Circuit Range/Performance
P0A02	Motor Electronics Coolant Temperature Sensor Circuit Low
P0A03	Motor Electronics Coolant Temperature Sensor Circuit High
P0A04	Motor Electronics Coolant Temperature Sensor Circuit Intermittent
P0A05	Motor Electronics Coolant Pump Control Circuit/Open
P0A06	Motor Electronics Coolant Pump Control Circuit Low
P0A07	Motor Electronics Coolant Pump Control Circuit High
P0A08	DC/DC Converter Status Circuit
P0A09	DC/DC Converter Status Circuit Low Input
P0A10	DC/DC Converter Status Circuit High Input
P0A11	DC/DC Converter Enable Circuit/Open
P0A12	DC/DC Converter Enable Circuit Low
P0A13	DC/DC Converter Enable Circuit High
P0A14	Engine Mount Control Circuit/Open
P0A15	Engine Mount Control Circuit Low
P0A16	Engine Mount Control Circuit High
P0A17	Motor Torque Sensor Circuit
P0A18	Motor Torque Sensor Circuit Range/Performance
P0A19	Motor Torque Sensor Circuit Low
P0A20	Motor Torque Sensor Circuit High
P0A21	Motor Torque Sensor Circuit Intermittent
P0A22	Generator Torque Sensor Circuit
P0A23	Generator Torque Sensor Circuit Range/Performance
P0A24	Generator Torque Sensor Circuit Low
P0A25	Generator Torque Sensor Circuit High
P0A26	Generator Torque Sensor Circuit Intermittent
P0A27	Battery Power Off Circuit
P0A28	Battery Power Off Circuit Low
P0A29	Battery Power Off Circuit High

Codes P0BXX, P0CXX, P0DXX, P0EXX, and P0FXX are reserved by SAE for future use.

P1XX DTCs are Reserved for OEM Use Only, as Follows:

P10XX	Manufacturer Controlled Fuel and Air Metering and Auxiliary Emission Controls
P11XX	Manufacturer Controlled Fuel and Air Metering
P12XX	Manufacturer Controlled Fuel and Air Metering
P13XX	Manufacturer Controlled Ignition System or Misfire
P14XX	Manufacturer Controlled Auxiliary Emission Controls
P15XX	Manufacturer Controlled Vehicle Speed, Idle Control, and Auxiliary Inputs
P16XX	Manufacturer Controlled Computer and Auxiliary Inputs
P17XX	Manufacturer Controlled Transmission
P18XX	Manufacturer Controlled Transmission
P19XX	Manufacturer Controlled Transmission

Designated categories for DTCs added in 2002 are, as follows:

P20XX Fuel and Air Metering and Auxiliary Emission Controls

DTC Number	DTC Naming
P2000	NO_x Trap Efficiency below Threshold, Bank 1
P2001	NO_x Trap Efficiency below Threshold, Bank 2
P2002	Particulate Trap Efficiency below Threshold, Bank 1
P2003	Particulate Trap Efficiency below Threshold, Bank 2
P2004	Intake Manifold Runner Control Stuck Open, Bank 1
P2005	Intake Manifold Runner Control Stuck Open, Bank 2
P2006	Intake Manifold Runner Control Stuck Closed, Bank 1
P2007	Intake Manifold Runner Control Stuck Closed, Bank 2
P2008	Intake Manifold Runner Control Circuit Open, Bank 1
P2009	Intake Manifold Runner Control Circuit Low, Bank 1
P2010	Intake Manifold Runner Control Circuit High, Bank 1
P2011	Intake Manifold Runner Control Circuit Open, Bank 2
P2012	Intake Manifold Runner Control Circuit Low, Bank 2
P2013	Intake Manifold Runner Control Circuit High, Bank 2
P2014	Intake Manifold Runner Position Sensor Switch Circuit, Bank 1
P2015	Intake Manifold Runner Position Sensor Switch Circuit Range/Performance, Bank 1
P2016	Intake Manifold Runner Position Sensor Switch Circuit Low, Bank 1
P2017	Intake Manifold Runner Position Sensor Switch Circuit High, Bank 1
P2018	Intake Manifold Runner Position Sensor Switch Circuit Intermittent
P2019	Intake Manifold Runner Position Sensor Switch Circuit, Bank 2
P2020	Intake Manifold Runner Position Sensor Switch Circuit Range/Performance, Bank 2
P2021	Intake Manifold Runner Position Sensor Switch Circuit Low, Bank 2
P2022	Intake Manifold Runner Position Sensor Switch Circuit High, Bank 2
P2023	Intake Manifold Runner Position Sensor Switch Circuit Intermittent, Bank 2
P2024	Evaporative Emissions (EVAP) Fuel Vapor Temperature Sensor Circuit
P2025	Evaporative Emissions (EVAP) Fuel Vapor Temperature Sensor Performance

DTC Number	DTC Naming
P2026	Evaporative Emissions (EVAP) Fuel Vapor Temperature Sensor Circuit Low Voltage
P2027	Evaporative Emissions (EVAP) Fuel Vapor Temperature Sensor Circuit High Voltage
P2028	Evaporative Emissions (EVAP) Fuel Vapor Temperature Sensor Circuit Intermittent
P2029	Fuel Fired Heater Disabled
P2030	Fuel Fired Heater Performance
P2031	Exhaust Gas Temperature Sensor Circuit, Bank 1 Sensor 2
P2032	Exhaust Gas Temperature Sensor Circuit Low, Bank 1 Sensor 2
P2033	Exhaust Gas Temperature Sensor Circuit High, Bank 1 Sensor 2
P2034	Exhaust Gas Temperature Sensor Circuit, Bank 2 Sensor 2
P2035	Exhaust Gas Temperature Sensor Circuit Low, Bank 2 Sensor 2
P2036	Exhaust Gas Temperature Sensor Circuit High, Bank 2 Sensor 2
P2037	Reductant Injection Air Pressure Sensor Circuit
P2038	Reductant Injection Air Pressure Sensor Circuit Range/Performance
P2039	Reductant Injection Air Pressure Sensor Circuit, Low Input
P2040	Reductant Injection Air Pressure Sensor Circuit, High Input
P2041	Reductant Injection Air Pressure Sensor Circuit, Intermittent
P2042	Reductant Temperature Sensor Circuit
P2043	Reductant Temperature Sensor Circuit Range/Performance
P2044	Reductant Temperature Sensor Circuit, Low Input
P2045	Reductant Temperature Sensor Circuit, High Input
P2046	Reductant Temperature Sensor Circuit, Intermittent
P2047	Reductant Injector Circuit Open, Bank 1 Unit 1
P2048	Reductant Injector Circuit Low, Bank 1 Unit 1
P2049	Reductant Injector Circuit High, Bank 1 Unit 1
P2050	Reductant Injector Circuit Open, Bank 2, Unit 1
P2051	Reductant Injector Circuit Low, Bank 2 Unit 1
P2052	Reductant Injector Circuit High, Bank 2 Unit 1
P2053	Reductant Injector Circuit Open, Bank 1 Unit 2
P2054	Reductant Injector Circuit Low, Bank 1 Unit 2
P2055	Reductant Injector Circuit High, Bank 1 Unit 2
P2056	Reductant Injector Circuit Open, Bank 2 Unit 2

DTC Number	DTC Naming
P2057	Reductant Injector Circuit Low, Bank 2 Unit 2
P2058	Reductant Injector Circuit High, Bank 2 Unit 2
P2059	Reductant Injection Air Pump Control Circuit Open
P2060	Reductant Injection Air Pump Control Circuit Low
P2061	Reductant Injection Air Pump Control Circuit High
P2062	Reductant Supply Control Circuit Open
P2063	Reductant Supply Control Circuit Low
P2064	Reductant Supply Control Circuit High
P2065	Fuel Level Sensor "B" Circuit
P2066	Fuel Level Sensor "B" Performance
P2067	Fuel Level Sensor "B" Low
P2068	Fuel Level Sensor "B" High
P2069	Fuel Level Sensor "B" Intermittent
P2070	Intake Manifold Tuning (IMT) Valve Stuck Open
P2071	Intake Manifold Tuning (IMT) Valve Stuck Closed
P2075	Intake Manifold Tuning (IMT) Valve Position Sensor, Switch Circuit
P2076	Intake Manifold Tuning (IMT) Valve Position Sensor, Switch Circuit Range/Performance
P2077	Intake Manifold Tuning (IMT) Valve Position Sensor, Switch Circuit Low
P2078	Intake Manifold Tuning (IMT) Valve Position Sensor, Switch Circuit High
P2079	Intake Manifold Tuning (IMT) Valve Position Sensor, Switch Circuit Intermittent
P2080	Exhaust Gas Temperature Sensor Circuit Range/Performance, Bank 1 Sensor 1
P2081	Exhaust Gas Temperature Sensor Circuit Intermittent, Bank 1 Sensor 1
P2082	Exhaust Gas Temperature Sensor Circuit Range/Performance, Bank 2 Sensor 1
P2083	Exhaust Gas Temperature Sensor Circuit Intermittent, Bank 2 Sensor 1
P2084	Exhaust Gas Temperature Sensor Circuit Range/Performance, Bank 1 Sensor 2
P2085	Exhaust Gas Temperature Sensor Circuit Intermittent, Bank 1 Sensor 2
P2086	Exhaust Gas Temperature Sensor Circuit Range/Performance, Bank 2 Sensor 2
P2087	Exhaust Gas Temperature Sensor Circuit Intermittent, Bank 2 Sensor 2

DTC Number	DTC Naming
P2088	"A" Camshaft Position Actuator Control Circuit Low, Bank 1
P2089	"A" Camshaft Position Actuator Control Circuit High, Bank 1
P2090	"B" Camshaft Position Actuator Control Circuit Low, Bank 1
P2091	"B" Camshaft Position Actuator Control Circuit High, Bank 1
P2092	"A" Camshaft Position Actuator Control Circuit Low, Bank 2
P2093	"A" Camshaft Position Actuator Control Circuit High, Bank 2
P2094	"B" Camshaft Position Actuator Control Circuit Low, Bank 2
P2095	"B" Camshaft Position Actuator Control Circuit High, Bank 2
P2096	Post Catalyst Fuel Trim System Too Lean, Bank 1
P2097	Post Catalyst Fuel Trim System Too Rich, Bank 1
P2098	Post Catalyst Fuel Trim System Too Lean, Bank 2
P2099	Post Catalyst Fuel Trim System Too Rich, Bank 2

P21XX Fuel and Air Metering and Auxiliary Emission Controls

DTC Number	DTC Naming
P2100	Throttle Actuator Control Motor Circuit Open
P2101	Throttle Actuator Control Motor Circuit Range/Performance
P2102	Throttle Actuator Control Motor Circuit Low
P2103	Throttle Actuator Control Motor Circuit High
P2104	Throttle Actuator Control System, Forced Idle
P2105	Throttle Actuator Control System, Forced Engine Shutdown
P2106	Throttle Actuator Control System, Forced Limited Power
P2107	Throttle Actuator Control Module Processor
P2108	Throttle Actuator Control Module Performance
P2109	Throttle/Pedal Position Sensor "A" Minimum Stop Performance
P2110	Throttle Actuator Control System, Forced Limited RPM
P2111	Throttle Actuator Control System Stuck Open
P2112	Throttle Actuator Control System Stuck Closed
P2113	Throttle/Pedal Position Sensor "B" Minimum Stop Performance
P2114	Throttle/Pedal Position Sensor "C" Minimum Stop Performance
P2115	Throttle/Pedal Position Sensor "D" Minimum Stop Performance
P2116	Throttle/Pedal Position Sensor "E" Minimum Stop Performance
P2117	Throttle/Pedal Position Sensor "F" Minimum Stop Performance
P2118	Throttle Actuator Control Motor Current Range/Performance
P2119	Throttle Actuator Control Throttle Body Range/Performance
P2120	Throttle/Pedal Position Sensor/Switch "D" Circuit
P2121	Throttle/Pedal Position Sensor/Switch "D" Circuit Range/Performance
P2122	Throttle/Pedal Position Sensor/Switch "D" Circuit Low Input
P2123	Throttle/Pedal Position Sensor/Switch "D" Circuit High Input
P2124	Throttle/Pedal Position Sensor/Switch "D" Circuit Intermittent
P2125	Throttle/Pedal Position Sensor/Switch "E" Circuit
P2126	Throttle/Pedal Position Sensor/Switch "E" Circuit Range/Performance
P2127	Throttle/Pedal Position Sensor/Switch "E" Circuit Low Input
P2128	Throttle/Pedal Position Sensor/Switch "E" Circuit High Input
P2129	Throttle/Pedal Position Sensor/Switch "E" Circuit Intermittent
P2130	Throttle/Pedal Position Sensor/Switch "F" Circuit

DTC Number	DTC Naming
P2131	Throttle/Pedal Position Sensor/Switch "F" Circuit Range/Performance
P2132	Throttle/Pedal Position Sensor/Switch "F" Circuit Low Input
P2133	Throttle/Pedal Position Sensor/Switch "F" Circuit High Input
P2134	Throttle/Pedal Position Sensor/Switch "F" Circuit Intermittent
P2135	Throttle/Pedal Position Sensor/Switch "A"/"B" Voltage Correlation
P2136	Throttle/Pedal Position Sensor/Switch "A"/"C" Voltage Correlation
P2137	Throttle/Pedal Position Sensor/Switch "B"/"C" Voltage Correlation
P2138	Throttle/Pedal Position Sensor/Switch "D"/"E" Voltage Correlation
P2139	Throttle/Pedal Position Sensor/Switch "D"/"F" Voltage Correlation
P2140	Throttle/Pedal Position Sensor/Switch "E"/"F" Voltage Correlation
P2141	Exhaust Gas Recirculation Throttle Control Circuit Low
P2142	Exhaust Gas Recirculation Throttle Control Circuit High
P2143	Exhaust Gas Recirculation Vent Control Circuit Open
P2144	Exhaust Gas Recirculation Vent Control Circuit Low
P2145	Exhaust Gas Recirculation Vent Control Circuit High
P2146	Fuel Injector Group "A" Supply Voltage Circuit, Open
P2147	Fuel Injector Group "A" Supply Voltage Circuit, Low
P2148	Fuel Injector Group "A" Supply Voltage Circuit, High
P2149	Fuel Injector Group "B" Supply Voltage Circuit, Open
P2150	Fuel Injector Group "B" Supply Voltage Circuit, Low
P2151	Fuel Injector Group "B" Supply Voltage Circuit, High
P2152	Fuel Injector Group "C" Supply Voltage Circuit, Open
P2153	Fuel Injector Group "C" Supply Voltage Circuit, Low
P2154	Fuel Injector Group "C" Supply Voltage Circuit, High
P2155	Fuel Injector Group "D" Supply Voltage Circuit, Open
P2156	Fuel Injector Group "D" Supply Voltage Circuit, Low
P2157	Fuel Injector Group "D" Supply Voltage Circuit, High
P2158	Vehicle Speed Sensor "B"
P2159	Vehicle Speed Sensor "B" Range/Performance
P2160	Vehicle Speed Sensor "B" Circuit Low
P2161	Vehicle Speed Sensor "B" Intermittent/Erratic
P2162	Vehicle Speed Sensor "A"/"B" Correlation

DTC Number	DTC Naming
P2163	Throttle/Pedal Position Sensor "A" Maximum Stop Performance
P2164	Throttle/Pedal Position Sensor "B" Maximum Stop Performance
P2165	Throttle/Pedal Position Sensor "C" Maximum Stop Performance
P2166	Throttle/Pedal Position Sensor "D" Maximum Stop Performance
P2167	Throttle/Pedal Position Sensor "E" Maximum Stop Performance
P2168	Throttle/Pedal Position Sensor "F" Maximum Stop Performance
P2169	Exhaust Pressure Regulator Vent Solenoid Control Circuit Open
P2170	Exhaust Pressure Regulator Vent Solenoid Control Circuit Low
P2171	Exhaust Pressure Regulator Vent Solenoid Control Circuit High
P2172	Throttle Actuator Control System, Sudden High Airflow Detected
P2173	Throttle Actuator Control System, High Airflow Detected
P2174	Throttle Actuator Control System, Sudden Low Airflow Detected
P2175	Throttle Actuator Control System, Low Airflow Detected
P2176	Throttle Actuator Control System, Idle Position Not Learned
P2177	System Too Lean Off Idle, Bank 1
P2178	System Too Rich Off Idle, Bank 1
P2179	System Too Lean Off Idle, Bank 2
P2180	System Too Rich Off Idle, Bank 2
P2181	Cooling System Performance
P2182	Engine Coolant Temperature Sensor 2 Circuit
P2183	Engine Coolant Temperature Sensor 2 Circuit Range/Performance
P2184	Engine Coolant Temperature Sensor 2 Circuit Low
P2185	Engine Coolant Temperature Sensor 2 Circuit High
P2186	Engine Coolant Temperature Sensor 2 Circuit Intermittent/Erratic
P2187	System Too Lean at Idle, Bank 1
P2188	System Too Rich at Idle, Bank 1
P2189	System Too Lean at Idle, Bank 2
P2190	System Too Rich at Idle, Bank 2
P2191	System Too Lean at Higher Load, Bank 1
P2192	System Too Rich at Higher Load, Bank 1
P2193	System Too Lean at Higher Load, Bank 2
P2194	System Too Rich at Higher Load, Bank 2

DTC Number	DTC Naming
P2195	O_2 Sensor Signal Stuck Lean, Bank 1, Sensor 1
P2196	O_2 Sensor Signal Stuck Rich, Bank 1, Sensor 1
P2197	O_2 Sensor Signal Stuck Lean, Bank 2, Sensor 1
P2198	O_2 Sensor Signal Stuck Rich, Bank 2, Sensor 1
P2199	Intake Air Temperature Sensor "1"/"2" Correlation

P22XX Fuel and Air Metering and Auxiliary Emission Controls

DTC Number	DTC Naming
P2200	NO_x Sensor Circuit, Bank 1
P2201	NO_x Sensor Circuit Range/Performance, Bank 1
P2202	NO_x Sensor Circuit Low Input, Bank 1
P2203	NO_x Sensor Circuit High Input, Bank 1
P2204	NO_x Sensor Circuit Intermittent Input, Bank 1
P2205	NO_x Sensor Heater Control Circuit Open, Bank 1
P2206	NO_x Sensor Heater Control Circuit Low, Bank 1
P2207	NO_x Sensor Heater Control Circuit High, Bank 1
P2208	NO_x Sensor Heater Sense Circuit, Bank 1
P2209	NO_x Sensor Heater Sense Circuit Range/Performance, Bank 1
P2210	NO_x Sensor Heater Sense Circuit Low, Bank 1
P2211	NO_x Sensor Heater Sense Circuit High, Bank 1
P2212	NO_x Sensor Heater Sense Circuit Intermittent, Bank 1
P2213	NO_x Sensor Circuit, Bank 2
P2214	NO_x Sensor Circuit Range/Performance, Bank 2
P2215	NO_x Sensor Circuit Low Input, Bank 2
P2216	NO_x Sensor Circuit High Input, Bank 2
P2217	NO_x Sensor Circuit Intermittent Input, Bank 2
P2218	NO_x Sensor Heater Control Circuit Open, Bank 2
P2219	NO_x Sensor Heater Control Circuit Low, Bank 2
P2220	NO_x Sensor Heater Control Circuit High, Bank 2
P2221	NO_x Sensor Heater Sense Circuit, Bank 2
P2222	NO_x Sensor Heater Sense Circuit Range/Performance, Bank 2
P2223	NO_x Sensor Heater Sense Circuit Low, Bank 2
P2224	NO_x Sensor Heater Sense Circuit High, Bank 2
P2225	NO_x Sensor Heater Sense Circuit Intermittent, Bank 2
P2226	Barometric Pressure Circuit
P2227	Barometric Pressure Circuit Range/Performance
P2228	Barometric Pressure Circuit Low
P2229	Barometric Pressure Circuit High
P2230	Barometric Pressure Circuit Intermittent

DTC Number	DTC Naming
P2231	O_2 Sensor Signal Circuit Shorted to Heater Circuit, Bank 1 Sensor 1
P2232	O_2 Sensor Signal Circuit Shorted to Heater Circuit, Bank 1 Sensor 2
P2233	O_2 Sensor Signal Circuit Shorted to Heater Circuit, Bank 1 Sensor 3
P2234	O_2 Sensor Signal Circuit Shorted to Heater Circuit, Bank 2 Sensor 1
P2235	O_2 Sensor Signal Circuit Shorted to Heater Circuit, Bank 2 Sensor 2
P2236	O_2 Sensor Signal Circuit Shorted to Heater Circuit, Bank 2 Sensor 3
P2237	O_2 Sensor Positive Current Control Circuit Open, Bank 1 Sensor 1
P2238	O_2 Sensor Positive Current Control Circuit Low, Bank 1 Sensor 1
P2239	O_2 Sensor Positive Current Control Circuit High, Bank 1 Sensor 1
P2240	O_2 Sensor Positive Current Control Circuit Open, Bank 2 Sensor 1
P2241	O_2 Sensor Positive Current Control Circuit Low, Bank 2 Sensor 1
P2242	O_2 Sensor Positive Current Control Circuit High, Bank 2 Sensor 1
P2243	O_2 Sensor Reference Voltage Circuit Open, Bank 1 Sensor 1
P2244	O_2 Sensor Reference Voltage Performance, Bank 1 Sensor 1
P2245	O_2 Sensor Reference Voltage Circuit Low, Bank 1 Sensor 1
P2246	O_2 Sensor Reference Voltage Circuit High, Bank 1 Sensor 1
P2247	O_2 Sensor Reference Voltage Circuit Open, Bank 2 Sensor 1
P2248	O_2 Sensor Reference Voltage Performance, Bank 2 Sensor 1
P2249	O_2 Sensor Reference Voltage Circuit Low, Bank 2 Sensor 1
P2250	O_2 Sensor Reference Voltage Circuit High, Bank 2 Sensor 1
P2251	O_2 Sensor Negative Current Control Circuit Open, Bank 1 Sensor 1
P2252	O_2 Sensor Negative Current Control Circuit Low, Bank 1 Sensor 1
P2253	O_2 Sensor Negative Current Control Circuit High, Bank 1 Sensor 1
P2254	O_2 Sensor Negative Current Control Circuit Open, Bank 2 Sensor 1
P2255	O_2 Sensor Negative Current Control Circuit Low, Bank 2 Sensor 1
P2256	O_2 Sensor Negative Current Control Circuit High, Bank 2 Sensor 1
P2257	Secondary Air Injection System Control "A" Circuit Low
P2258	Secondary Air Injection System Control "A" Circuit High
P2259	Secondary Air Injection System Control "B" Circuit Low
P2260	Secondary Air Injection System Control "B" Circuit High
P2261	Turbo/Supercharger Bypass Valve, Mechanical
P2262	Turbo Boost Pressure Not Detected, Mechanical

DTC Number	DTC Naming
P2263	Turbo/Supercharger Boost System Performance
P2264	Water in Fuel Sensor Circuit
P2265	Water in Fuel Sensor Circuit Range/Performance
P2266	Water in Fuel Sensor Circuit Low
P2267	Water in Fuel Sensor Circuit High
P2268	Water in Fuel Sensor Circuit Intermittent
P2269	Water in Fuel Condition
P2270	O_2 Sensor Signal Stuck Lean, Bank 1 Sensor 2
P2271	O_2 Sensor Signal Stuck Rich, Bank 1 Sensor 2
P2272	O_2 Sensor Signal Stuck Lean, Bank 2 Sensor 2
P2273	O_2 Sensor Signal Stuck Rich, Bank 2 Sensor 2
P2274	O_2 Sensor Signal Stuck Lean, Bank 1 Sensor 3
P2275	O_2 Sensor Signal Stuck Rich, Bank 1 Sensor 3
P2276	O_2 Sensor Signal Stuck Lean, Bank 2 Sensor 3
P2277	O_2 Sensor Signal Stuck Rich, Bank 2 Sensor 3
P2278	O_2 Sensor Signals Swapped, Bank 1 Sensor 3/Bank 2 Sensor 3
P2279	Intake Air System Leak
P2280	Airflow Restriction/Air Leak Between Air Filter and MAF Sensor
P2281	Air Leak Between MAF Sensor and Throttle Body
P2282	Air Leak Between Throttle Body and Intake Valves
P2283	Injector Control Pressure Sensor Circuit
P2284	Injector Control Pressure Sensor Circuit Range/Performance
P2285	Injector Control Pressure Sensor Circuit Low
P2286	Injector Control Pressure Sensor Circuit High
P2287	Injector Control Pressure Sensor Circuit Intermittent
P2288	Injector Control Pressure Too High
P2289	Injector Control Pressure Too High, Engine Off
P2290	Injector Control Pressure Too Low
P2291	Injector Control Pressure Too Low, Engine Cranking
P2292	Injector Control Pressure Erratic
P2293	Fuel Pressure Regulator 2, Performance
P2294	Fuel Pressure Regulator 2 Control Circuit

DTC Number	DTC Naming
P2295	Fuel Pressure Regulator 2 Control Circuit Low
P2296	Fuel Pressure Regulator 2 Control Circuit High
P2297	O_2 Sensor Out of Range During Deceleration, Bank 1 Sensor 1
P2298	O_2 Sensor Out of Range During Deceleration, Bank 2 Sensor 1
P2299	Brake Pedal Position/Accelerator Pedal Position Incompatible

P23XX Ignition System or Misfire

DTC Number	DTC Naming
P2300	Ignition Coil "A" Primary Control Circuit Low
P2301	Ignition Coil "A" Primary Control Circuit High
P2302	Ignition Coil "A" Secondary Circuit
P2303	Ignition Coil "B" Primary Control Circuit Low
P2304	Ignition Coil "B" Primary Control Circuit High
P2305	Ignition Coil "B" Secondary Circuit
P2306	Ignition Coil "C" Primary Control Circuit Low
P2307	Ignition Coil "C" Primary Control Circuit High
P2308	Ignition Coil "C" Secondary Circuit
P2309	Ignition Coil "D" Primary Control Circuit Low
P2310	Ignition Coil "D" Primary Control Circuit High
P2311	Ignition Coil "D" Secondary Circuit
P2312	Ignition Coil "E" Primary Control Circuit Low
P2313	Ignition Coil "E" Primary Control Circuit High
P2314	Ignition Coil "E" Secondary Circuit
P2315	Ignition Coil "F" Primary Control Circuit Low
P2316	Ignition Coil "F" Primary Control Circuit High
P2317	Ignition Coil "F" Secondary Circuit
P2318	Ignition Coil "G" Primary Control Circuit Low
P2319	Ignition Coil "G" Primary Control Circuit High
P2320	Ignition Coil "G" Secondary Circuit
P2321	Ignition Coil "H" Primary Control Circuit Low
P2322	Ignition Coil "H" Primary Control Circuit High
P2323	Ignition Coil "H" Secondary Circuit
P2324	Ignition Coil "I" Primary Control Circuit Low
P2325	Ignition Coil "I" Primary Control Circuit High
P2326	Ignition Coil "I" Secondary Circuit
P2327	Ignition Coil "J" Primary Control Circuit Low
P2328	Ignition Coil "J" Primary Control Circuit High
P2329	Ignition Coil "J" Secondary Circuit
P2330	Ignition Coil "K" Primary Control Circuit Low

DTC Number	DTC Naming
P2331	Ignition Coil "K" Primary Control Circuit High
P2332	Ignition Coil "K" Secondary Circuit
P2333	Ignition Coil "L" Primary Control Circuit Low
P2334	Ignition Coil "L" Primary Control Circuit High
P2335	Ignition Coil "L" Secondary Circuit
P2336	Cylinder #1 above Knock Threshold
P2337	Cylinder #2 above Knock Threshold
P2338	Cylinder #3 above Knock Threshold
P2339	Cylinder #4 above Knock Threshold
P2340	Cylinder #5 above Knock Threshold
P2341	Cylinder #6 above Knock Threshold
P2342	Cylinder #7 above Knock Threshold
P2343	Cylinder #8 above Knock Threshold
P2344	Cylinder #9 above Knock Threshold
P2345	Cylinder #10 above Knock Threshold
P2346	Cylinder #11 above Knock Threshold
P2347	Cylinder #12 above Knock Threshold

P24XX Auxiliary Emission Controls

DTC Number	DTC Naming
P2400	Evaporative Emission System Leak Detection Pump Control Circuit Open
P2401	Evaporative Emission System Leak Detection Pump Control Circuit Low
P2402	Evaporative Emission System Leak Detection Pump Control Circuit High
P2403	Evaporative Emission System Leak Detection Pump Sense Circuit Open
P2404	Evaporative Emission System Leak Detection Pump Sense Circuit Range/Performance
P2405	Evaporative Emission System Leak Detection Pump Sense Circuit Low
P2406	Evaporative Emission System Leak Detection Pump Sense Circuit High
P2407	Evaporative Emission System Leak Detection Pump Sense Circuit Intermittent/Erratic
P2408	Fuel Cap Sensor Switch Circuit
P2409	Fuel Cap Sensor Switch Circuit Range/Performance
P2410	Fuel Cap Sensor Switch Circuit Low
P2411	Fuel Cap Sensor Switch Circuit High
P2412	Fuel Cap Sensor Switch Circuit Intermittent/Erratic
P2413	Exhaust Gas Recirculation System Performance
P2414	O_2 Sensor Exhaust Sample Error, Bank 1 Sensor 1
P2415	O_2 Sensor Exhaust Sample Error, Bank 2 Sensor 1
P2416	O_2 Sensor Signals Swapped, Bank 1 Sensor 2/Bank 1 Sensor 3
P2417	O_2 Sensor Signals Swapped, Bank 2 Sensor 2/Bank 2 Sensor 3
P2418	Evaporative Emission System Switching Valve Control Circuit Open
P2419	Evaporative Emission System Switching Valve Control Circuit Low
P2420	Evaporative Emission System Switching Valve Control Circuit High
P2421	Evaporative Emission System Vent Valve Stuck Open
P2422	Evaporative Emission System Vent Valve Stuck Closed
P2423	HC Adsorption Catalyst Efficiency below Threshold, Bank 1
P2424	HC Adsorption Catalyst Efficiency below Threshold, Bank 2
P2425	Exhaust Gas Recirculation Cooling Valve Control Circuit Open
P2426	Exhaust Gas Recirculation Cooling Valve Control Circuit Low
P2427	Exhaust Gas Recirculation Cooling Valve Control Circuit High
P2428	Exhaust Gas Temperature Too High, Bank 1

DTC Number	DTC Naming
P2429	Exhaust Gas Temperature Too High, Bank 2
P2430	Secondary Air Injection System Airflow/Pressure Sensor Circuit, Bank 1
P2431	Secondary Air Injection System Airflow/Pressure Sensor Circuit Range/Performance, Bank 1
P2432	Secondary Air Injection System Airflow/Pressure Sensor Circuit Low, Bank 1
P2433	Secondary Air Injection System Airflow/Pressure Sensor Circuit High, Bank 1
P2434	Secondary Air Injection System Airflow/Pressure Sensor Circuit Intermittent/Erratic, Bank 1
P2435	Secondary Air Injection System Airflow/Pressure Sensor Circuit, Bank 2
P2436	Secondary Air Injection System Airflow/Pressure Sensor Circuit Range/Performance, Bank 2
P2437	Secondary Air Injection System Airflow/Pressure Sensor Circuit Low, Bank 2
P2438	Secondary Air Injection System Airflow/Pressure Sensor Circuit High, Bank 2
P2439	Secondary Air Injection System Airflow/Pressure Sensor Circuit Intermittent/Erratic, Bank 2
P2440	Secondary Air Injection System Switching Valve Stuck Open, Bank 1
P2441	Secondary Air Injection System Switching Valve Stuck Closed, Bank 1
P2442	Secondary Air Injection System Switching Valve Stuck Open, Bank 2
P2443	Secondary Air Injection System Switching Valve Stuck Closed, Bank 2
P2444	Secondary Air Injection System Pump Stuck On, Bank 1
P2445	Secondary Air Injection System Pump Stuck Off, Bank 1
P2446	Secondary Air Injection System Pump Stuck On, Bank 2
P2447	Secondary Air Injection System Pump Stuck Off, Bank 2

P25XX Auxiliary Inputs

DTC Number	DTC Naming
P2500	Generator Lamp/L Terminal Circuit Low
P2501	Generator Lamp/L Terminal Circuit High
P2502	Charging System Voltage
P2503	Charging System Voltage Low
P2504	Charging System Voltage High
P2505	ECM/PCM Power Input Signal
P2506	ECM/PCM Power Input Signal Range/Performance
P2507	ECM/PCM Power Input Signal Low
P2508	ECM/PCM Power Input Signal High
P2509	ECM/PCM Power Input Signal Intermittent
P2510	ECM/PCM Power Relay Sense Circuit Range/Performance
P2511	ECM/PCM Power Relay Sense Circuit Intermittent
P2512	Event Data Recorder Request Circuit Open
P2513	Event Data Recorder Request Circuit Low
P2514	Event Data Recorder Request Circuit High
P2515	A/C Refrigerant Pressure Sensor "B" Circuit
P2516	A/C Refrigerant Pressure Sensor "B" Circuit Range/Performance
P2517	A/C Refrigerant Pressure Sensor "B" Circuit Low
P2518	A/C Refrigerant Pressure Sensor "B" Circuit High
P2519	A/C Request "A" Circuit
P2520	A/C Request "A" Circuit Low
P2521	A/C Request "A" Circuit High
P2522	A/C Request "B" Circuit
P2523	A/C Request "B" Circuit Low
P2524	A/C Request "B" Circuit High
P2525	Vacuum Reservoir Pressure Sensor Circuit
P2526	Vacuum Reservoir Pressure Sensor Circuit Range/Performance
P2527	Vacuum Reservoir Pressure Sensor Circuit Low
P2528	Vacuum Reservoir Pressure Sensor Circuit High
P2529	Vacuum Reservoir Pressure Sensor Circuit Intermittent
P2530	Ignition Run Position Circuit

DTC Number	DTC Naming
P2531	Ignition Run Position Circuit Low
P2532	Ignition Run Position Circuit High
P2533	Ignition Run/Start Position Circuit
P2534	Ignition Run/Start Position Circuit Low
P2535	Ignition Run/Start Position Circuit High
P2536	Ignition Accessory Position Circuit
P2537	Ignition Accessory Position Circuit Low
P2538	Ignition Accessory Position Circuit High
P2539	Low Pressure Fuel System Sensor Circuit
P2540	Low Pressure Fuel System Sensor Circuit Range/Performance
P2541	Low Pressure Fuel System Sensor Circuit Low
P2542	Low Pressure Fuel System Sensor Circuit High
P2543	Low Pressure Fuel System Sensor Circuit Intermittent
P2544	Torque Management Request Input Signal "A"
P2545	Torque Management Request Input Signal "A" Range/Performance
P2546	Torque Management Request Input Signal "A" Low
P2547	Torque Management Request Input Signal "A" High
P2548	Torque Management Request Input Signal "B"
P2549	Torque Management Request Input Signal "B" Range/Performance
P2550	Torque Management Request Input Signal "B" Low
P2551	Torque Management Request Input Signal "B" High
P2552	Throttle/Fuel Inhibit Circuit
P2553	Throttle/Fuel Inhibit Circuit Range/Performance
P2554	Throttle/Fuel Inhibit Circuit Low
P2555	Throttle/Fuel Inhibit Circuit High
P2556	Engine Coolant Level Sensor/Switch Circuit
P2557	Engine Coolant Level Sensor/Switch Circuit Range/Performance
P2558	Engine Coolant Level Sensor/Switch Circuit Low
P2559	Engine Coolant Level Sensor/Switch Circuit High
P2560	Engine Coolant Level Low
P2561	A/C Control Module Requested MIL Illumination
P2562	Turbocharger Boost Control Position Sensor Circuit

DTC Number	DTC Naming
P2563	Turbocharger Boost Control Position Sensor Circuit Range/Performance
P2564	Turbocharger Boost Control Position Sensor Circuit Low
P2565	Turbocharger Boost Control Position Sensor Circuit High
P2566	Turbocharger Boost Control Position Sensor Circuit Intermittent
P2567	Direct Ozone Reduction Catalyst Temperature Sensor Circuit
P2568	Direct Ozone Reduction Catalyst Temperature Sensor Circuit Range/Performance
P2569	Direct Ozone Reduction Catalyst Temperature Sensor Circuit Low
P2570	Direct Ozone Reduction Catalyst Temperature Sensor Circuit High
P2571	Direct Ozone Reduction Catalyst Temperature Sensor Circuit Intermittent/Erratic
P2572	Direct Ozone Reduction Catalyst Deterioration Sensor Circuit
P2573	Direct Ozone Reduction Catalyst Deterioration Sensor Circuit Range/Performance
P2574	Direct Ozone Reduction Catalyst Deterioration Sensor Circuit Low
P2575	Direct Ozone Reduction Catalyst Deterioration Sensor Circuit High
P2576	Direct Ozone Reduction Catalyst Deterioration Sensor Circuit Intermittent/Erratic
P2577	Direct Ozone Reduction Catalyst Efficiency below Threshold

P26XX Computer and Auxiliary Outputs

DTC Number	DTC Naming
P2600	Coolant Pump Control Circuit Open
P2601	Coolant Pump Control Circuit Range/Performance
P2602	Coolant Pump Control Circuit Low
P2603	Coolant Pump Control Circuit High
P2604	Intake Air Heater "A" Circuit Range/Performance
P2605	Intake Air Heater "A" Circuit Open
P2606	Intake Air Heater "B" Circuit Range/Performance
P2607	Intake Air Heater "B" Circuit Low
P2608	Intake Air Heater "B" Circuit High
P2609	Intake Air Heater System Performance
P2610	ECM/PCM Internal Engine Off Timer Performance
P2611	A/C Refrigerant Distribution Valve Control Circuit Open
P2612	A/C Refrigerant Distribution Valve Control Circuit Low
P2613	A/C Refrigerant Distribution Valve Control Circuit High
P2614	Camshaft Position Signal Output Circuit Open
P2615	Camshaft Position Signal Output Circuit Low
P2616	Camshaft Position Signal Output Circuit High
P2617	Crankshaft Position Signal Output Circuit Open
P2618	Crankshaft Position Signal Output Circuit Low
P2619	Crankshaft Position Signal Output Circuit High
P2620	Throttle Position Signal Output Circuit Open
P2621	Throttle Position Signal Output Circuit Low
P2622	Throttle Position Signal Output Circuit High
P2623	Injector Control Pressure Regulator Circuit Open
P2624	Injector Control Pressure Regulator Circuit Low
P2625	Injector Control Pressure Regulator Circuit High
P2626	O_2 Sensor Pumping Current Trim Circuit Open, Bank 1 Sensor 1
P2627	O_2 Sensor Pumping Current Trim Circuit Low, Bank 1 Sensor 1
P2628	O_2 Sensor Pumping Current Trim Circuit High, Bank 1 Sensor 1
P2629	O_2 Sensor Pumping Current Trim Circuit Open, Bank 2 Sensor 1
P2630	O_2 Sensor Pumping Current Trim Circuit Low, Bank 2 Sensor 1

DTC Number	DTC Naming
P2631	O_2 Sensor Pumping Current Trim Circuit High, Bank 2 Sensor 1
P2632	Fuel Pump "B" Control Circuit Open
P2633	Fuel Pump "B" Control Circuit Low
P2634	Fuel Pump "B" Control Circuit High
P2635	Fuel Pump "A" Low Flow/ Performance
P2636	Fuel Pump "B" Low Flow/Performance
P2637	Torque Management Feedback Signal "A"
P2638	Torque Management Feedback Signal "A" Range/Performance
P2639	Torque Management Feedback Signal "A" Low
P2640	Torque Management Feedback Signal "A" High
P2641	Torque Management Feedback Signal "B"
P2642	Torque Management Feedback Signal "B" Range/Performance
P2643	Torque Management Feedback Signal "B" Low
P2644	Torque Management Feedback Signal "B" High
P2645	"A" Rocker Arm Actuator Control Circuit Open, Bank 1
P2646	"A" Rocker Arm Actuator System Performance or Stuck Off, Bank 1
P2647	"A" Rocker Arm Actuator System Stuck On, Bank 1
P2648	"A" Rocker Arm Actuator Control Circuit Low, Bank 1
P2649	"A" Rocker Arm Actuator Control Circuit High, Bank 1
P2650	"B" Rocker Arm Actuator Control Circuit Open, Bank 1
P2651	"B" Rocker Arm Actuator System Performance or Stuck Off, Bank 1
P2652	"B" Rocker Arm Actuator System Stuck On, Bank 1
P2653	"B" Rocker Arm Actuator Control Circuit Low, Bank 1
P2654	"B" Rocker Arm Actuator Control Circuit High, Bank 1
P2655	"A" Rocker Arm Actuator Control Circuit Open, Bank 2
P2656	"A" Rocker Arm Actuator System Performance or Stuck Off, Bank 2
P2657	"A" Rocker Arm Actuator System Stuck On, Bank 2
P2658	"A" Rocker Arm Actuator Control Circuit Low, Bank 2
P2659	"A" Rocker Arm Actuator Control Circuit High, Bank 2
P2660	"B" Rocker Arm Actuator Control Circuit Open, Bank 2
P2661	"B" Rocker Arm Actuator System Performance or Stuck Off, Bank 2
P2662	"B" Rocker Arm Actuator System Stuck On, Bank 2

DTC Number	DTC Naming
P2663	"B" Rocker Arm Actuator Control Circuit Low, Bank 2
P2664	"B" Rocker Arm Actuator Control Circuit High, Bank 2
P2665	Fuel Shutoff Valve "B" Control Circuit Open
P2666	Fuel Shutoff Valve "B" Control Circuit Low
P2667	Fuel Shutoff Valve "B" Control Circuit High
P2668	Fuel Mode Indicator Lamp Control Circuit
P2669	Actuator Supply Voltage "B" Circuit Open
P2670	Actuator Supply Voltage "B" Circuit Low
P2671	Actuator Supply Voltage "B" Circuit High

P27XX Transmission

DTC Number	DTC Naming
P2700	Transmission Function Element "A" Apply Time Range/Performance
P2701	Transmission Function Element "B" Apply Time Range/Performance
P2702	Transmission Function Element "C" Apply Time Range/Performance
P2703	Transmission Function Element "D" Apply Time Range/Performance
P2704	Transmission Function Element "E" Apply Time Range/Performance
P2705	Transmission Function Element "F" Apply Time Range/Performance
P2706	Shift Solenoid "F"
P2707	Shift Solenoid "F" Performance or Stuck Off
P2708	Shift Solenoid "F" Stuck On
P2709	Shift Solenoid "F" Electrical
P2710	Shift Solenoid "F" Intermittent
P2711	Unexpected Mechanical Gear Disengagement
P2712	Hydraulic Power Unit Leakage
P2713	Pressure Control Solenoid "D"
P2714	Pressure Control Solenoid "D" Performance or Stuck Off
P2715	Pressure Control Solenoid "D" Stuck On
P2716	Pressure Control Solenoid "D" Electrical
P2717	Pressure Control Solenoid "D" Intermittent
P2718	Pressure Control Solenoid "D" Control Circuit Open
P2719	Pressure Control Solenoid "D" Control Circuit Range/Performance
P2720	Pressure Control Solenoid "D" Control Circuit Low
P2721	Pressure Control Solenoid "D" Control Circuit High
P2722	Pressure Control Solenoid "E"
P2723	Pressure Control Solenoid "E" Performance or Stuck Off
P2724	Pressure Control Solenoid "E" Stuck On
P2725	Pressure Control Solenoid "E" Electrical
P2726	Pressure Control Solenoid "E" Intermittent
P2727	Pressure Control Solenoid "E" Control Circuit Open
P2728	Pressure Control Solenoid "E" Control Circuit Range/Performance
P2729	Pressure Control Solenoid "E" Control Circuit Low
P2730	Pressure Control Solenoid "E" Control Circuit High

DTC Number	DTC Naming
P2731	Pressure Control Solenoid "F"
P2732	Pressure Control Solenoid "F" Performance or Stuck Off
P2733	Pressure Control Solenoid "F" Stuck On
P2734	Pressure Control Solenoid "F" Electrical
P2735	Pressure Control Solenoid "F" Intermittent
P2736	Pressure Control Solenoid "F" Control Circuit Open
P2737	Pressure Control Solenoid "F" Control Circuit Range/Performance
P2738	Pressure Control Solenoid "F" Control Circuit Low
P2739	Pressure Control Solenoid "F" Control Circuit High
P2740	Transmission Fluid Temperature Sensor "B" Circuit
P2741	Transmission Fluid Temperature Sensor "B" Circuit Range/Performance
P2742	Transmission Fluid Temperature Sensor "B" Circuit Low
P2743	Transmission Fluid Temperature Sensor "B" Circuit High
P2744	Transmission Fluid Temperature Sensor "B" Circuit Intermittent
P2745	Intermediate Shaft Speed Sensor "B" Circuit
P2746	Intermediate Shaft Speed Sensor "B" Circuit Range/Performance
P2747	Intermediate Shaft Speed Sensor "B" Circuit No Signal
P2748	Intermediate Shaft Speed Sensor "B" Circuit Intermittent
P2749	Intermediate Shaft Speed Sensor "C" Circuit
P2750	Intermediate Shaft Speed Sensor "C" Circuit Range/Performance
P2751	Intermediate Shaft Speed Sensor "C" Circuit No Signal
P2752	Intermediate Shaft Speed Sensor "C" Circuit Intermittent
P2753	Transmission Fluid Cooler Control Circuit Open
P2754	Transmission Fluid Cooler Control Circuit Low
P2755	Transmission Fluid Cooler Control Circuit High
P2756	Torque Converter Clutch Pressure Control Solenoid
P2757	Torque Converter Clutch Pressure Control Solenoid Control Circuit Performance or Stuck Off
P2758	Torque Converter Clutch Pressure Control Solenoid Control Circuit Stuck On
P2759	Torque Converter Clutch Pressure Control Solenoid Control Circuit Electrical
P2760	Torque Converter Clutch Pressure Control Solenoid Control Circuit Intermittent

DTC Number	DTC Naming
P2761	Torque Converter Clutch Pressure Control Solenoid Control Circuit Open
P2762	Torque Converter Clutch Pressure Control Solenoid Control Circuit Range/Performance
P2763	Torque Converter Clutch Pressure Control Solenoid Control Circuit High
P2764	Torque Converter Clutch Pressure Control Solenoid Control Circuit Low
P2765	Input/Turbine Speed Sensor "B" Circuit
P2766	Input/Turbine Speed Sensor "B" Circuit Range/Performance
P2767	Input/Turbine Speed Sensor "B" Circuit No Signal
P2768	Input/Turbine Speed Sensor "B" Circuit Intermittent
P2769	Torque Converter Clutch Circuit Low
P2770	Torque Converter Clutch Circuit High
P2771	Four-Wheel Drive (4WD) Low Switch Circuit
P2772	Four-Wheel Drive (4WD) Low Switch Circuit Range/Performance
P2773	Four-Wheel Drive (4WD) Low Switch Circuit Low
P2774	Four-Wheel Drive (4WD) Low Switch Circuit High
P2775	Upshift Switch Circuit Range/Performance
P2776	Upshift Switch Circuit Low
P2777	Upshift Switch Circuit High
P2778	Upshift Switch Circuit Intermittent/Erratic
P2779	Downshift Switch Circuit Range/Performance
P2780	Downshift Switch Circuit Low
P2781	Downshift Switch Circuit High
P2782	Downshift Switch Circuit Intermittent/Erratic
P2783	Torque Converter Temperature Too High
P2784	Input/Turbine Speed Sensor "A"/"B" Correlation
P2785	Clutch Actuator Temperature Too High
P2786	Gear Shift Actuator Temperature Too High
P2787	Clutch Temperature Too High
P2788	Auto Shift Manual Adaptive Learning at Limit
P2789	Clutch Adaptive Learning at Limit
P2790	Gate Select Direction Circuit

DTC Number	DTC Naming
P2791	Gate Select Direction Circuit Low
P2792	Gate Select Direction Circuit High
P2793	Gear Shift Direction Circuit
P2794	Gear Shift Direction Circuit Low
P2795	Gear Shift Direction Circuit High

P028XX ISO/SAE Reserved

P2AXX Fuel and Air Metering and Auxiliary Emission Controls

DTC Number	DTC Naming
P2A00	O_2 Sensor Range/Performance, Bank 1 Sensor 1
P2A01	O_2 Sensor Range/Performance, Bank 1 Sensor 2
P2A02	O_2 Sensor Range/Performance, Bank 1 Sensor 3
P2A03	O_2 Sensor Range/Performance, Bank 2 Sensor 1
P2A04	O_2 Sensor Range/Performance, Bank 2, Sensor 2
P2A05	O_2 Sensor Range/Performance, Bank 2, Sensor 3

P30XX Fuel and Air Metering and Auxiliary Emission Controls

P31XX Fuel and Air Metering and Auxiliary Emission Controls

P32XX Fuel and Air Metering and Auxiliary Emission Controls

P33XX Ignition System of Misfire

P34XX Cylinder Deactivation

DTC Number	DTC Naming
P3400	Cylinder Deactivation system, Bank 1
P3401	Cylinder 1 Deactivation/Intake Valve Control Circuit Open
P3402	Cylinder 1 Deactivation/Intake Valve Control Performance
P3403	Cylinder 1 Deactivation/Intake Valve Control Circuit Low
P3404	Cylinder 1 Deactivation/Intake Valve Control Circuit High
P3405	Cylinder 1 Exhaust Valve Control Circuit Open
P3406	Cylinder 1 Exhaust Valve Control Performance
P3407	Cylinder 1 Exhaust Valve Control Circuit Low
P3408	Cylinder 1 Exhaust Valve Control Circuit High
P3409	Cylinder 2 Deactivation/Intake Valve Control Circuit Open
P3410	Cylinder 2 Deactivation/Intake Valve Control Performance
P3411	Cylinder 2 Deactivation/Intake Valve Control Circuit Low
P3412	Cylinder 2 Deactivation/Intake Valve Control Circuit High
P3413	Cylinder 2 Exhaust Valve Control Circuit Open
P3414	Cylinder 2 Exhaust Valve Control Performance
P3415	Cylinder 2 Exhaust Valve Control Circuit Low
P3416	Cylinder 2 Exhaust Valve Control Circuit High
P3417	Cylinder 3 Deactivation/Intake Valve Control Circuit Open
P3418	Cylinder 3 Deactivation/Intake Valve Control Performance
P3419	Cylinder 3 Deactivation/Intake Valve Control Circuit Low
P3420	Cylinder 3 Deactivation/Intake Valve Control Circuit High
P3421	Cylinder 3 Exhaust Valve Control Circuit Open
P3422	Cylinder 3 Exhaust Valve Control Performance
P3423	Cylinder 3 Exhaust Valve Control Circuit Low
P3424	Cylinder 3 Exhaust Valve Control Circuit High
P3425	Cylinder 4 Deactivation/Intake Valve Control Circuit Open
P3426	Cylinder 4 Deactivation/Intake Valve Control Performance
P3427	Cylinder 4 Deactivation/Intake Valve Control Circuit Low
P3428	Cylinder 4 Deactivation/Intake Valve Control Circuit High
P3429	Cylinder 4 Exhaust Valve Control Circuit Open
P3430	Cylinder 4 Exhaust Valve Control Performance

DTC Number	DTC Naming
P3431	Cylinder 4 Exhaust Valve Control Circuit Low
P3432	Cylinder 4 Exhaust Valve Control Circuit High
P3433	Cylinder 5 Deactivation/Intake Valve Control Circuit Open
P3434	Cylinder 5 Deactivation/Intake Valve Control Performance
P3435	Cylinder 5 Deactivation/Intake Valve Control Circuit Low
P3436	Cylinder 5 Deactivation/Intake Valve Control Circuit High
P3437	Cylinder 5 Exhaust Valve Control Circuit Open
P3438	Cylinder 5 Exhaust Valve Control Performance
P3439	Cylinder 5 Exhaust Valve Control Circuit Low
P3440	Cylinder 5 Exhaust Valve Control Circuit High
P3441	Cylinder 6 Deactivation/Intake Valve Control Circuit Open
P3442	Cylinder 6 Deactivation/Intake Valve Control Performance
P3443	Cylinder 6 Deactivation/Intake Valve Control Circuit Low
P3444	Cylinder 6 Deactivation/Intake Valve Control Circuit High
P3445	Cylinder 6 Exhaust Valve Control Circuit Open
P3446	Cylinder 6 Exhaust Valve Control Performance
P3447	Cylinder 6 Exhaust Valve Control Circuit Low
P3448	Cylinder 6 Exhaust Valve Control Circuit High
P3449	Cylinder 7 Deactivation/Intake Valve Control Circuit Open
P3450	Cylinder 7 Deactivation/Intake Valve Control Performance
P3451	Cylinder 7 Deactivation/Intake Valve Control Circuit Low
P3452	Cylinder 7 Deactivation/Intake Valve Control Circuit High
P3453	Cylinder 7 Exhaust Valve Control Circuit Open
P3454	Cylinder 7 Exhaust Valve Control Performance
P3455	Cylinder 7 Exhaust Valve Control Circuit Low
P3456	Cylinder 7 Exhaust Valve Control Circuit High
P3457	Cylinder 8 Deactivation/Intake Valve Control Circuit Open
P3458	Cylinder 8 Deactivation/Intake Valve Control Performance
P3459	Cylinder 8 Deactivation/Intake Valve Control Circuit Low
P3460	Cylinder 8 Deactivation/Intake Valve Control Circuit High
P3461	Cylinder 8 Exhaust Valve Control Circuit Open
P3462	Cylinder 8 Exhaust Valve Control Performance

DTC Number	DTC Naming
P3463	Cylinder 8 Exhaust Valve Control Circuit Low
P3464	Cylinder 8 Exhaust Valve Control Circuit High
P3465	Cylinder 9 Deactivation/Intake Valve Control Circuit Open
P3466	Cylinder 9 Deactivation/Intake Valve Control Performance
P3467	Cylinder 9 Deactivation/Intake Valve Control Circuit Low
P3468	Cylinder 9 Deactivation/Intake Valve Control Circuit High
P3469	Cylinder 9 Exhaust Valve Control Circuit Open
P3470	Cylinder 9 Exhaust Valve Control Performance
P3471	Cylinder 9 Exhaust Valve Control Circuit Low
P3472	Cylinder 9 Exhaust Valve Control Circuit High
P3473	Cylinder 10 Deactivation/Intake Valve Control Circuit Open
P3474	Cylinder 10 Deactivation/Intake Valve Control Performance
P3475	Cylinder 10 Deactivation/Intake Valve Control Circuit Low
P3476	Cylinder 10 Deactivation/Intake Valve Control Circuit High
P3477	Cylinder 10 Exhaust Valve Control Circuit Open
P3478	Cylinder 10 Exhaust Valve Control Performance
P3479	Cylinder 10 Exhaust Valve Control Circuit Low
P3480	Cylinder 10 Exhaust Valve Control Circuit High
P3481	Cylinder 11 Deactivation/Intake Valve Control Circuit Open
P3482	Cylinder 11 Deactivation/Intake Valve Control Performance
P3483	Cylinder 11 Deactivation/Intake Valve Control Circuit Low
P3484	Cylinder 11 Deactivation/Intake Valve Control Circuit High
P3485	Cylinder 11 Exhaust Valve Control Circuit Open
P3486	Cylinder 11 Exhaust Valve Control Performance
P3487	Cylinder 11 Exhaust Valve Control Circuit Low
P3488	Cylinder 11 Exhaust Valve Control Circuit High
P3489	Cylinder 12 Deactivation/Intake Valve Control Circuit Open
P3490	Cylinder 12 Deactivation/Intake Valve Control Performance
P3491	Cylinder 12 Deactivation/Intake Valve Control Circuit Low
P3492	Cylinder 12 Deactivation/Intake Valve Control Circuit High
P3493	Cylinder 12 Exhaust Valve Control Circuit Open
P3494	Cylinder 12 Exhaust Valve Control Performance

DTC Number	DTC Naming
P3495	Cylinder 12 Exhaust Valve Control Circuit Low
P3496	Cylinder 12 Exhaust Valve Control Circuit High
P3497	Cylinder Deactivation System, Bank 2

P35XX ISO/SAE Reserved

P36XX ISO/SAE Reserved

P37XX ISO/SAE Reserved

P38XX ISO/SAE Reserved

P39XX ISO/SAE Reserved

APPENDIX D: Nonstandard OBD II DLC Locations through 2000 Model Year

The following chart illustrates typical nonstandard, but allowed, locations for the DLC. Most scan tools or diagnostic software will state the location of the DLC as data for the vehicle year, make, and model are entered.

Manufacturer	Model(s)	Model Year(s)	DLC Location
Acura	CL	1997–1998	Under dash, passenger side near center console
Acura	CL	1999	Uncovered, above shifter
Acura	RL	1999–2000	Center console, forward of shifter, behind cover
Acura	TL	1996–1998	Center console, behind ashtray
Acura	TL	1999–2000	Behind center dash/console, below stereo, near seat heater control at left
Acura	Integra	1996–1999	Under dash, passenger side near center console
Acura	NSX, S2000	1999–2000	Under dash, passenger side near center console
Acura	RL	1996–1998	Front of center console, passenger side
Audi	A4, A4 Avant, Cabriolet	1996	Center console, behind rear sliding ashtray cover
Audi	A6	1996–1997	Center console, behind front tray
Bentley	All	1996–2000	In glove box, behind cover

Manufacturer	Model(s)	Model Year(s)	DLC Location
BMW	3 Series, 5 series, M3	1996–2000	Behind left side of lower left dash, but covered by panel. Turn slotted screw 1/4 turn to open.
BMW	7 Series	1996–2000	Behind center dash/console, under stereo controls
BMW	X3, M Roadster	1996–2000	Behind passenger side of center dash/console
BMW	Z3	1996–2000	Behind cover, under dash on passenger side
Ferrari	All	1996–2000	Very high under dash, driver's side near center of car
Ford	Bronco, F-Series trucks	1996	Under dash, slightly right of center, covered
Ford	Thunderbird	1996–97	Under dash, slightly right of center, covered
Honda	Accord	1996–97	Behind ashtray, center console
Honda	CR-V, Prelude	1997–2000	Under dash, passenger side near center
Honda	Del Sol, Insight	1996–2000	Under dash, passenger side near center
Honda	Odyssey	1996–98	Behind passenger side of center dash/console
Honda	Prelude	1996	Uncovered, above shifter
Hyundai	Accent	1996–98	Right center dash, in coin holder
Land Rover	Defender	1997	Left center of dash, under tray
Land Rover	Range Rover	1996–2000	Under right dash, behind cover
Lotus	Espirit	1997–2000	Under cover, above right center of dash/console
Porsche	All	1996	Behind center dash, toward left side
Rolls-Royce	All	1996–2000	In glove box, behind cover
Toyota	Prius	2000	Behind right center dash/console
Toyota	Previa	1996–97	Behind cover, right side of instrument cluster
Volvo	850	1997–98	Center console, behind coin holder, forward of shifter
Volvo	All except 850 and S80	1998–99	Behind right side of center console, near hand brake

Manufacturer	Model(s)	Model Year(s)	DLC Location
Volvo	S40, V40	2000	Behind cover, left center dash/console
Volvo	C70, S70, V70	2000	Behind cover, center console forward of shifter
Volkswagen	Cabrio, Golf, Jetta	1996–98	Behind right center dash, to right of ashtray
Volkswagen	Eurovan	1996–early 99	Under cover, right of instrument cluster, behind wiper control lever
Volkswagen	Golf, Jetta	1999	Behind right center dash
Volkswagen	Passat	1996–97	Under cover, right of instrument cluster, behind wiper control lever

Index